Mary Lawton
230 C
Highl[illegible] [illegible], Ill.

CODICES SELECTI

Die Reihe dient der Erforschung und Erschließung bedeutender Handschriften durch vollständige Faksimileausgaben. Sie wurde begonnen und fortgesetzt mit Unterstützung der nachstehenden Bibliotheken und ihrer Leiter, deren Namen chronologisch nach dem Erscheinen der einzelnen Bände aufgeführt sind.

Österreichische Nationalbibliothek - Josef Stummvoll, Generaldirektor emeritus
und Rudolf Fiedler, Generaldirektor
Biblioteca Capitolare, Verona - Giuseppe Turrini, Prefetto
Museum für Völkerkunde, Wien - Frau Etta Becker-Donner †, Direktor
British Museum, London - Sir Frank Francis K. C. B., Director emeritus
and Principal Librarian
Museo de América, Madrid - Señora Doña Pilar F. Vega de Ferrandis, Directora
Bibliothèque nationale, Paris - Etienne Dennery, Administrateur Général
Kärntner Landesarchiv, Klagenfurt - Gotbert Moro, Direktor
Bodleian Library, Oxford - J. N. L. Myres, Bodleys Librarian
Biblioteca Apostolica Vaticana - Alfonso Raes, Prefetto emerito
e Alfons M. Stickler, Prefetto
Biblioteca Universitaria, Bologna - Guglielmo Manfrè, Direttore emerito
e Luigia Risoldi, Direttrice
Universitätsbibliothek, Würzburg - Josef Tiwisina, Direktor
Biblioteca Nazionale Centrale, Firenze - Emanuele Casamassima, Direttore emer.
e Anna Maria Giorgetti-Vichi, Direttrice
City of Liverpool Museum - T. A. Hume, Director
Tiroler Landesmuseum Ferdinandeum, Innsbruck - Erich Egg, Direktor
Kupferstichkabinett der Staatlichen Museen Preussischer Kulturbesitz, Berlin
Matthias Winner, Direktor
Benediktinerstift Kremsmünster - Albert Bruckmayr O. S. B., Abt
Topkapı Sarayı Müzesi, Istanbul - Kemal Çig, Direktor
Staatsbibliothek Preussischer Kulturbesitz - Ludwig Borngässer, Generaldir. emer.
und Ekkehart Vesper, Generaldirektor
Bibliothèque de l'Assemblée Nationale, Paris - J. M. Priou, Bibliothécaire en Chef
Stadtbibliothek Trier - Richard Laufner, Direktor
Benediktinerstift St. Peter, Salzburg - Erzabt Franz Bachler O. S. B.
Bibliothek des Museums für angewandte Kunst, Wien - Gerhart Egger, Direktor
Sächsische Landesbibliothek, Dresden - Burghart Burgemeister, Direktor
The Cleveland Museum of Art - Sherman E. Lee, Director

WISSENSCHAFTLICHE BERATER UND KOMMENTATOREN

Franz Unterkircher, Wien - Otto Mazal, Wien - Franz Sauer, Graz
Charles Gibson, Iowa - Karl Anton Nowotny, Köln - Otto Adelhofer, Wien
Cottie A. Burland, London - Ferdinand Anders, Wien - Alfred Kracher, Graz
Hans Thurn, Würzburg - Hans Gerstinger †, Graz - Gerhard Schmidt, Wien
Antoine De Schryver, Gent - Carl A. Willemsen, Bonn - Hedwig Heger, Wien
Alfons M. Stickler, Rom - Walter Höflechner, Graz - Charles H. Talbot, London
Florentine Mütherich, München - Norbert Richard Wolf, Innsbruck
Hans Butzmann, Wolfenbüttel - Peter Bloch, Berlin - Kurt Holter, Wels
Willibrord Neumüller O. S. B., Kremsmünster - Vjekoslav Štefanić,
Biserka Grabar, Anica Nazor, Marija Pantelić, Zagreb - Marcel Thomas, Paris
Werner Abraham, Groningen - Wolfgang Hilger, Wien - Otto Demus, Wien
Reiner Haussherr, Bonn - Bernhard Bischoff, München
Jacqueline de Durand-Forest, Paris - Felix Vongrey O. Cist., Lilienfeld
Herbert Douteil, Köln - Richard Laufner, Trier - Peter K. Klein, Bonn
Hanns Swarzenski, Boston - Karl Forstner, Salzburg
Le Duc de Brissac, Paris - François Avril, Paris - Helmut Deckert, Dresden
Pramod Chandra, Chicago

CODICES SELECTI

PHOTOTYPICE IMPRESSI

FACSIMILE VOL. LV
COMMENTARIUM VOL. LV*

AKADEMISCHE DRUCK- u. VERLAGSANSTALT
GRAZ – AUSTRIA
1976

ṬŪṬĪ-NĀMA

TALES OF A PARROT – DAS PAPAGEIENBUCH

VOLLSTÄNDIGE FAKSIMILE-AUSGABE IM ORIGINALFORMAT
DER HANDSCHRIFT AUS DEM BESITZ DES CLEVELAND MUSEUM OF ART

COMPLETE COLOUR FACSIMILE EDITION IN ORIGINAL SIZE
OF THE MANUSCRIPT IN POSSESSION OF THE CLEVELAND MUSEUM OF ART

COMMENTARIUM

PRAMOD CHANDRA

AKADEMISCHE DRUCK- u. VERLAGSANSTALT
GRAZ – AUSTRIA
1976

Gedruckt mit Unterstützung des Cleveland Museum of Art

Für die Blätter 190r, 190v, 205r, 205v, 212r, 212v, 235r, 238r, 238v, 247r, 247v, 286r, 286v aus Privatbesitz wurden die fotografischen Unterlagen vom Cleveland Museum of Art zur Verfügung gestellt.

Gesamtherstellung: Akademische Druck- u. Verlagsanstalt

Printed in Austria
ISBN 3-201-00959-8
94.76

PRAMOD CHANDRA

THE ṬŪṬĪ-NĀMA

of The Cleveland Museum of Art and the Origins of Mughal Painting

Foreword by Sherman E. Lee

To my parents

CONTENTS

FOREWORD

Discoveries, real ones, are the usually unattainable dreams of scholars and collectors alike. They are only too human, and the excitement of discovery would seem to be a basic human emotion—so much so that immeasurable quantities of time, energy, and money have been spent on both successful and unsuccessful explorations in all fields where one hopes the unknown waits just beyond the horizon. The discovery of works of art should really be called rediscovery, for the object is man-made for a purpose and usually for a patron. The artist is the original discoverer—or creator, which is one and the same—but history records numerous works, important in their own day, lost for decades or centuries, only to be rediscovered and take their place in the continuity of the history of art. When the rediscovery is accompanied by the opportunity to save the work from destruction or disfiguration, one can only call it a double act of Fortune, today the most underrated of ancient deities.

Since the modern history of the *Ṭūṭī-nāma* is one of these doubly fortunate rediscoveries, it seems only fitting to give as close and factual account of the circumstances as possible. Dr. Chandra's text provides, in extended and complete form, the art-historical context and critical evaluation of the manuscript. The considerable critical attention given the *Ṭūṭī-nāma* attests to its important position in the history of Mughal painting and the analysis and arguments given here reaffirm and amplify those given in the preliminary publication of the manuscript in *The Burlington Magazine* of 1963. To this significant accomplishment we now append the modern history of the Cleveland *Ṭūṭī-nāma.*

In early October, 1962, a mimeographed sheet addressed to the Director's Office arrived at The Cleveland Museum of Art. It offered for sale individual leaves of an eighteenth-century Persian manuscript at the derisory price of fifteen dollars per page. Clipped to the form letter were Kodachrome prints of five such leaves, now represented by folios 215, 267, 293b, 305, and 307 of the recovered manuscript. Cursory examination of the photographs revealed that the pages in question were neither Persian nor eighteenth century. What little I knew of Mughal painting had been added to by a pilgrimage to Vienna two years previously where I had been able to see for the first time the well-known monumental pages of the *Ḥamza-nāma,* Humayun's and Akbar's great commission. Even the crude Kodachromes showed gold, hot oranges, purples, and mustard yellow (unlike hues of Persian origin) as well as aggressive representations of rocks and ornament closely analogous to those depicted in the *Ḥamza-nāma.* The fish of the swirling rivers as well as the realistic inventiveness of the man beating the ass were obviously Mughal and early. What was to be done?

A telephone call to the sender, a Bernard Brown Agency in Milwaukee, extracted the information that the miniatures in question were being razorcut from a complete manuscript and being offered about at numerous colleges and universities in the East and Middle West. Under the circumstances of willful despoliation of a presumably intact manuscript, I offered to purchase this "Persian" book because of its "subject interest," provided the dispersed pages could be recovered. Fortunately the necessary funds were generously provided by Mrs. A. Dean Perry, a long-time friend of the Museum. Then began three weeks of slow frustration. Some pages were at a photographer's; others were out on approval with various agents. The manuscript, minus some thirty to forty pages, arrived in mid-October with its green leather Victorian binding (Fig. 1). Other pages arrived in the mail from various sources over a period of weeks. Ultimately all but 13 folios were recovered, and 7 of these were sold to private collectors and are included in the present publication. The richness and complexity of the material was overwhelming, for it became increasingly clear that many different hands and styles were represented in the miniatures. What had seemed to be a Hamza-influenced manuscript now was clearly a far more subtle and complicated problem, involving disparate pre-Hamza schools of painting.

Further inquiry made it possible to partially trace the modern history of the manuscript. Brown had obtained it from the Philadelphia antique dealer Harry Burke, who in turn had purchased it at public auction. This sale had been held on Wednesday, October 7, 1959, at the house of William D. Morley Inc., Chestnut Street, Philadelphia, and comprised part of the estate of the Hon. Breckinridge Long of Laurel, Maryland. Mr. Long had been a United States State Department official and presumably had acquired the manuscript abroad. It figured as item 2303: "Persian Illuminated Manuscript. Tooti Nameh, or Tales of a Parrot. With 215 illuminated paintings by a native artist. Circa 1760. Full dark green morocco and gilt by Riviere." The sale price was $ 300.00! The level of accuracy of the sale catalog can be judged also from the next entry, number 2304: Whistler's "General Art of Making Enemies."

Pencil notes on the inside covers and flyleaves gave the following information: "Manuscript, Persian, circa 1760, Tooti Nameh, or Tales of a Parrot, upwards of 300 leaves with 215 illuminated paintings by a native artist-maaxx." "Tootee Namuh." "noco bg $\frac{\text{aoso}}{\text{shoes}}$." "Purchased Sot [?] (erased) Price 260£ 1902." Presumably the "Sot . . ." reference is to a Sotheby sale of 1902, but search to date has not revealed the specific reference.

Since almost 1/5th of the pages had already been razored out and the binding was of the late 19th century, it was decided to dismantle the rest of the manuscript so

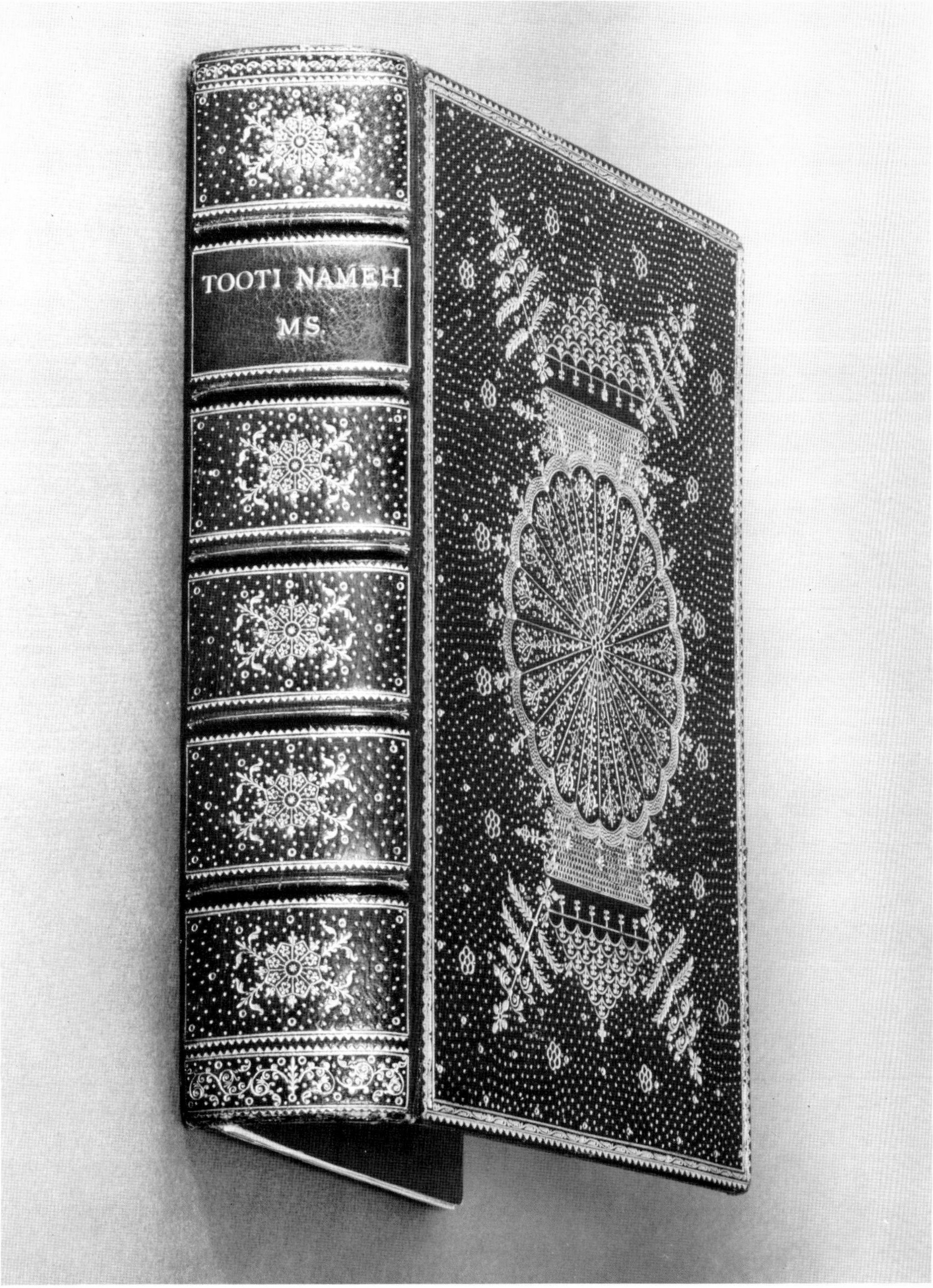

Fig. 1

that the leaves could be preserved and shown individually. Many of the yellows, oranges, and whites had become somewhat oxidized. A careful and closely supervised treatment with peroxide vapour was effective in removing much of the oxidization and restoring the original appearance of the colors.

There remained the collation, study, and publication of the manuscript. A preliminary publication was made as soon as possible *(The Burlington Magazine)*, but a careful and complete publication has been delayed until now. Mrs. Margaret Marcus, now retired from the staff of the Museum, did much of the earliest studies of the manuscript, including collation. Dr. Pramod Chandra, who agreed with me very early as to the seminal importance of this *Ṭūṭī-nāma,* happily and fortunately went on with the study of the illustrations. The results of his research are well worth the time expended. Dr. Muhammed Simsar, retired from the Library of Congress, became fascinated with the literary importance of the text and agreed to make a full translation and commentary which is being published separately from this facsimile edition but designed to be used with it. Dr. Stanislaw Czuma, now Curator of Indian Art at The Cleveland Museum of Art, has given invaluable help in the later stages of research and publication. I should also like to thank Dr. Stuart C. Welch of the Fogg Art Museum, Harvard University, and William G. Archer, retired Keeper of Indian Art at the Victoria and Albert Museum, for their continued encouragement. New discoveries often make their way against heavy opposition. The faith and dedication of all those named, and many others, to the *Ṭūṭī-nāma* so fortunately recovered and now preserved at Cleveland, is heartening in itself, but particularly because it resulted in this full and handsome publication by the Akademische Druck- u. Verlagsanstalt.

Sherman E. Lee

PREFACE

Sometime in 1963, over a decade ago, Dr. Sherman E. Lee communicated to me news of the acquisition by the Cleveland Museum of Art of the *Ṭūṭī-nāma*, and I still recall the sense of excitement with which I first saw the manuscript. It seemed too good to be true. Not only did it contain some wonderful miniatures of the most refreshingly beautiful quality, but also a great deal of clear evidence that showed the Indian sources of the Mughal style, only hints of which had been available previously. As it was felt by us to be important to acquaint our colleagues immediately with what seemed a vital discovery, Dr. Lee and I rapidly prepared a preliminary study of the manuscript for the *Burlington Magazine* issue of December 1963; and I was also entrusted with the task of presenting a paper at the XXVIth International Congress of Orientalists which was held in New Delhi that year. It was our view that the manuscript represented the most archaic and formative phases of Akbar's atelier, actually demonstrating the processes by which the work of non-Mughal artists was transformed into the Mughal style proper. As is to be expected when startlingly fresh discoveries of this type are announced, there was a reluctance on the part of some scholars to accept our opinion, and at least one was disconcerted enough to promptly suggest the possibility of a forgery. A more sober but nevertheless animated discussion, largely unpublished, took place in scholarly circles, and surfaced at a seminar on Mughal painting organized by me at The American Academy of Benares in 1967. Dr. Ananda Krishna's recent paper on the subject is essentially an enlargement and restatement of the views which he expressed on that occasion and which evoked the most lively debate by other participants, notably Mr. Robert Skelton, Dr. Moti Chandra, Mr. Ashok Das, Rai Krishnadasa, and myself.

Meanwhile a comprehensive publication on the manuscript, a long time in the planning and, for one reason or the other, considerably delayed, has only now been completed, over ten years after the initial publication. Originally, Dr. Lee and I had hoped to work jointly on the manuscript; but his manifold duties made this impossible and it was left to me to carry on the work as best I could. I have, however, constantly kept in touch with Dr. Lee, and I am very grateful not only for his providing me the opportunity to work on the manuscript, but also for the several stimulating discussions I have had with him over the course of the years. My colleagues Mr. Stuart C. Welch of the Fogg Art Museum and Mr. Robert Skelton of the Victoria and Albert Museum have been helpful as always with problems, and I am deeply indebted to them for their generous advice, friendship and support, as I am to my father Dr. Moti Chandra and Mr. W. G. Archer who constantly encourage me in my work.

For help with problems involving a knowledge of Persian, I am especially grateful to Professor C. M. Naim of the University of Chicago who translated the relevant sections of the Persian texts which are included in Appendices A–D. He also spent countless hours in deciphering and translating the various inscriptions, and has been ever ready to discuss lucidly and patiently whatever problems I chose to confront him with. Dr. Z. A. Desai, the Superintending Epigraphist for Arabic and Persian Inscriptions, Archaeological Survey of India, Maulana Imtiyaz Ali Arshi and Akbar Ali Khan Arshizadah of the Raza Library, Rampur, have also been of considerable assistance. I cannot thank them enough for the readiness and promptness of their answers to my several queries.

The transliteration of Persian words, suggested by Professor Naim, has been devised to indicate the orthography and to some extent the pronunciation of Persian as was written and spoken in India. A word of explanation also seems to be necessary with regard to the transliteration of Sanskrit and Hindi words as they appear in Persian texts, these being consistently rendered as though they were written in the Devanāgarī alphabet. Thus, we have Basāvana for the more usual Basāwan; and though this may be initially unfamiliar, it has the advantage of representing the word as it was written in the native alphabet, and not as it was approximated in Persian. It is hoped that this will contribute to the standardisation of Roman spellings and all the clarity and convenience this provides. Hindi and Sanskrit words have been rendered into Roman script according to the international system of transliteration, the short vowel *a* being always represented.

The above acknowledgements by no means exhaust the numerous friends and scholars who have helped me in one way or the other. I should like to particularly record my thanks to Dr. H. Bojer of the Bayerische Staatsbibliothek, Munich for help in making the important *Nafāʾis al-Maʾāṣir* manuscript available; Dr. Stanislaw Czuma and Mrs. Margaret F. Marcus of the Cleveland Museum of Art for help in studying the *Ṭūṭī-nāma* manuscript, and for solving problems connected with the illustration of this book; Dr. Mehmed A. Simsar, whose accompanying translation of the *Ṭūṭī-nāma* is a pleasure to read; Dr. Ratan Parimoo and Mr. Michael Naumer for providing some important photographs; and to my students and friends Mr. Charles W. Ervin and Miss Marsha Tajima who have often given invaluable assistance and who have also been the testing ground for many of my thoughts and ideas.

The University of Chicago — P. C.
September, 1974

INTRODUCTION

The Mughal style of painting which began in the reign of Akbar (1556–1605) and lasted for almost three hundred years was the most important of Indian styles of its age. In its flourishing days, which covered the reigns of Akbar, Jahāngīr (1605–28) and, to a lesser extent, that of Shāh Jahān (1628–58), its achievements were particularly distinguished, constituting a renaissance which made its influence felt all over India. The style, inaugurated with the 1400 overwhelming illustrations to the *Qiṣṣa-i Amīr Ḥamza,* gives the appearance of having been fully formed at birth, leading early investigators to conclude that it was of Persian origin, the result of artists and artistic influences emanating from that country. This conclusion, however, was never wholly satisfactory, for the style had an originality, vitality, and an inner unity that could not be explained by foreign influences alone. Perhaps the tastes of the patrons had something to do with its distinctiveness, for though the early Mughals were themselves foreigners, the emperor Akbar had ceased to consider himself one. The ever present "Indian environment" could also be suggested as the reason for the individuality of the style; but the works of art produced in this environment were of a totally different character, and the connection they may have had with the Mughal style seemed incapable of demonstration. True, the more knowledge was gained regarding pre-Akbar Indian styles, the more plausible they became as one of the sources accounting for the distinctive character of Mughal painting; but conclusive visual proof was still lacking. The illustrated manuscript of the *Ṭūṭī-nāma* in the Cleveland Museum of Art which forms the subject of this study provides the visual testimony for the connection between previously existing Indian schools and Mughal painting, and also demonstrates how artists trained in indigenous idioms were transformed into practitioners of the Mughal style.

In order to fully understand the part played by this unique manuscript in the origin and early development of Mughal painting, it was found necessary to place it in the proper context with respect to both the non-Mughal Indian schools and the early Mughal style and tradition. We therefore propose, at the outset, to give a brief account of what little we know of painting and patronage in the reign of Bābar and Humāyūn, followed with an account, admittedly as yet quite sketchy, of non-Mughal Indian painting traditions current during their times and earlier (ca. 1450–1550) and then a description of the Mughal style proper as it flourished early in the reign of Akbar, its founder. The appropriate context being thus provided, we will proceed to give a fairly detailed account of the Cleveland

Ṭūṭī-nāma, for only then will it become possible for us to appreciate its correct place in the origin and evolution of Mughal painting.

In adopting this approach, we may sometimes give the impression of wandering afield; but this has, in a sense, been imposed on us by the paucity of adequate publications on the subject. This, together with the limited and uncertain nature of the knowledge available to scholars of Indian painting of this period, has made it necessary to sketch, in some instances for the first time, preliminary outlines of styles of painting intimately connected with our study of the *Ṭūṭī-ṇāma*. Without doing this, it would be impossible to come to grips with the fascinating problems presented by this crucial manuscript.

CHAPTER I
PAINTING AND PATRONAGE UNDER BĀBAR AND HUMĀYŪN

Bābar and the Timurid tradition

Bābar (born 1483, died 1530), the first Mughal emperor of India, was among the last of the Timurid princes, the distinguished house which ruled over most of Central Asia and Iran from 1370, the year of Tīmūr's accession to the throne, upto the opening years of the sixteenth century.[1] Even as he established his great empire in India after being dislodged from his ancestral principality of Ferghana, and even as other Timurid rulers collapsed around him with inexorable regularity, he remembered the most glorious time of his life to be the short period when he had captured and ruled the central Timurid capital of Samarkand.[2] The successors of Timur, who did so much to repair the ravages of their warlike ancestor, were among the most cultured rulers of the age, patrons of literature, the arts and the sciences, and often greatly accomplished in their own persons. Bābar was no exception; rather he was among the most outstanding.[3] His memoirs are a masterpiece of Turki prose literature, his poetry is accomplished; and though not a single painting produced for him has been found, it is proper to notice in this study of Mughal painting what little we know of his personal perception of the arts. This should give us some idea of the kind of cultural milieu and attitudes of mind prevalent in his circle, the same being inherited presumably by those of his successors in India who laid the foundation of Mughal painting.

We do not know if Bābar, who wrote such exquisite prose, himself practiced painting, though he was a fine calligrapher, having invented a type of writing called the Khaṭṭ-i Bābarī.[4] It would be no surprise, however, if he did. It was a common enough tradition among members of the ruling houses of Central Asia to be involved deeply enough in art to attempt to practice it themselves.[5] Bābar himself draws

1. Bābar ruled Ferghana from 1494–1502, Kabul and its dependancies from 1504 to 1526, and sat on the throne of Delhi from 1526 to 1530.
2. *Bābur-nāma,* p. 134.
3. "In fact no one in his family before him ever possessed such talents as him," says Mirzā Muḥammad Ḥaidar, his cultivated cousin, *Tarikh-i-Rashidi,* p. 174.
4. *Bābur-nāma,* Appendix Q, p., lxii,ff.
5. Yūnus Khān of Tashkent (ob. 1487), Bābar's maternal grandfather, for example, was skilled in painting as was Shaibānī Khān, Bābar's inveterate enemy, who is reproached with ordering Bihzād, "Draw in this manner." See *Tarikh-i-Rashidi,* p. 155 and the *Tuḥfa-i Sāmī,* a work written by Sām Mirzā, Shāh Ṭahmāsp's brother, in 1550–51, quoted in Mahfuz-ul Haq, p. 241.

attention to painting being part of the accomplishment of princes. Thus, a cousin Bāʾisung͟hur, son of Sulṭān Maḥmūd of Samarkand, is eulogised as an excellent calligrapher and "in painting also his hand was not bad;" and Mirzā Muḥammad Ḥaidar, another cousin of whom we will hear later, is described as "possessing a hand deft in everything, penmanship and painting."[6]

Whether Bābar painted or not, he had decided opinions on artists and patrons. He praises S͟hāh Muẓaffar as a painter of "dainty portraits, representing the hair very daintily." Bihzād's work he also considered to be dainty, "but he did not draw beardless faces well; he used greatly to lengthen the double chin; bearded faces he drew admirably; and it was through instruction of their patron Mīr ʿAlī S͟her Nawāʾī that they became distinguished in painting."[7] Whether or not what Bābar has to say is correct is hardly relevant, but what is revealing is the intellectual consciousness in which such a remark has its origin, demonstrating a clear conception of the positive role of the patron in the making of a work of art. If Bābar had painters, it is easy to imagine their being moulded by his taste and personality very much as the painters of his grandson, the emperor Akbar, indeed were.

Humāyūn and Mirzā Muḥammad Ḥaidar

The brave and compassionate Humāyūn, brought up lovingly by his father, must have shared his interest in art; but, though we begin to get paintings dating from his reign, we do not get the kind of forthright statement that so clearly revealed Bābar's thoughts on the matter. Humāyūn was an accomplished calligrapher;[8] and that he was fond of beautifully written and illustrated manuscripts is also clear. His sorrow at the theft in Cambay of a precious *Tīmūr-nāma* of Mullā Hātifī illustrated by Bihzād is recorded,[9] as is his great pleasure at recovering the royal books lost in the confusion of the battle of Kipchak in 1550 when he had stationed himself in Kabul.[10] These two bits of information also testify to books being considered indispensable enough adjuncts of life so as to accompany the emperor even during arduous military campaigns.

Though we do not have any directly expressed views of Humāyūn on the subject, some idea of the kind of thoughts entertained about painters and painting during his time may be obtained from the remarks of Mirzā Muḥammad Ḥaidar (ca. 1500—

---

6. *Bābur-nāma*, pp. 111, 22.
7. *Ibid.*, pp. 291, 272.
8. According to Sām Mirzā, his handwriting was perfect. See Mahfuz-ul Haq, p. 243.
9. See Chaghatai, *Painting During the Sultanate Period*, p. 37, fn. 3. The incident must have taken place during the Gujarat campaign of 1535—6 against Bahādur S͟hāh, several years before the Persian exile.
10. *Akbar Nāmā* I, p. 57.

1551), the same cousin whose deftness as a painter had been alreadv noticed by Bābar, and who later entered the service of Humāyūn.[11]

Mirzā Ḥaidar confirms Bābar's views of his (Ḥaidar's) accomplishments as a painter by stating that in painting and illuminating he was a past master.[12] But he was not just that, his remarks on the qualities of the great painters of his time revealing him to be a true connoisseur and discerning critic possessed of sensitive and individualistic judgement as well.[13] Speaking of a painter named Manṣūr, he praises his extremely fine brushwork, the strokes being firmer and therefore more refreshing *(khunuktar)* than those of his son Shāh Muẕaffar who however surpassed him in other respects by virtue of "an exceedingly delicate *(nāzukī)* brush, so clear *(ṣāfi)*, refined *(malāḥat)* and matured *(pokhtagī)* that the eyes of all beholders were amazed." Mirzā Muḥammad Ḥaidar has his own views of the universally admired Bihzād as well, declaring that though the preliminary design *(ṭarḥ)* and composition of his pictures was superior to that of Shāh Muẕaffar, and his brush firmer, it did not have the latter's delicacy. Bihzād's finish was even better than that of his master Maulānā Mīrak Naqqāsh, though his designs *(ṭarḥ)* were less mature. In rank, Mirzā Ḥaidar gave the first place to Khwāja ʿAbd al-Ḥayy, who lived in an earlier period under the *khāqāns* of the house of Hulāgū, followed by Shāh Muẕaffar and Bihzād, like whom there had been no other painters right up to the time of the author's writing. He too, like Bābar, does not forget to mention that both Shāh Muẕaffar and Bihzād were patronised by Mīr ʿAlī Sher (ob. 1501). The works of Qāsim ʿAli, the portrait painter and pupil of Bihzād, approximated those of his teacher, but in this style *(uslūb)*, as "any expert connoisseur can recognize," his works are rougher *(durushttar)*, the original designs more unbalanced *(bī andāmtar)*. Bābā Ḥājī was unequalled in Khurasan in sketching design and drawing in charcoal, and was reputed to have once drawn "fifty circles and a half" all exactly alike, as though

11. The mothers of Mirzā Ḥaidar and Bābar were sisters, being daughters of Yūnus Khān. Mirzā Ḥaidar was born in A. H. 905/1499–1500, joined Sulṭān Abū Saʿid Khān of Kashghar, and in 1531 undertook an extraordinary expedition to Ladakh, Kashmir, Baltistan, and Tibet with the intention of destroying the temple at Lhasa. He failed, and in 1536 we find him in the service of Mirzā Kāmrān who ruled Kabul, Kandahar, and the Punjab under the nominal suzerainity of his brother, the emperor Humāyūn. When Kāmrān abandoned his brother, Mirzā Ḥaidar joined Humāyūn inspite of the latter's desperate situation. He commanded the Mughal troops in the fateful Battle of Bilgram (May 1540), and during Humāyūn's withdrawal, himself conquered Kashmir. When Humāyūn conquered Kabul in 1551, Mirzā Ḥaidar acknowledged Humāyūn's sovereignity, striking coins in his name. He died in 1551. He wrote the *Tārīkh-i Rashīdī*, his celebrated work on history, while ruler of Kashmir. The account of Mirzā Ḥaidar is abstracted from Elias and Ross, *A History of the Moghuls of Central Asia.*
12. *Tarikh-i-Rashidi*, p. 4.
13. The discussion which follows is largely based on Arnold's translation of the relevant passages as reproduced in BWG, pp. 189–91. For a sensitive exposition of the technical terms see Schroeder, p. 10.

made by a pair of compasses. Mirzā Ḥaidar's own teacher Darwi̱sh Muḥammad[14] was capable of a tour de force also, drawing on a single grain of rice a man on horseback lifting a lion on the point of a javelin. He was a pupil of Sh̲āh Muẕaffar, and, though in fineness of brush he had no equal, even surpassing his master, he was not so "symmetrical or expert or refined," and apt to make very crude strokes. Mullā Yūsuf was a pupil of Bihzād who worked thrice as fast as other masters, though his brush was not "as agreeable, his gilding being superior to his painting"[15].

Mirzā Ḥaidar was also a painter, and this was not very unusual, for many princes practiced the art; but his discerning and knowledgeable discourse on the art and his judgements on the style of painters are exceptional, reflecting to some extent, surely, the artistic ideals that must have been prevalent at Humāyūn's court, a court to which he was related both by ties of blood and service.

Humāyūn in Persia

We do not know of Humāyūn's achievements as a painter, though there is some evidence to prove that he learned painting like other princes of his time;[16] nor do we know of his tastes except from the few surviving works of the artists he later gathered around him. Judging, however, from his actions in Persia, he must have had a strong predilection for the art, making earnest attempts, even in ill fortune, to recruit the best painters. In this task, he seems to have been quite successful, his personal presence in the country and Sh̲āh Ṭahmāsp's disenchantment with art proving to be most helpful.

After his final defeat at the hands of Sh̲er Sh̲āh in the Battle of Bilgram (1540), Humāyūn was forced to flee, the years 1541 and 1542 being spent wandering in Sind and Rajasthan where the future emperor Akbar was born in 1542. Finally entering Persia via Garamsir in December 1543–January 1544, and welcomed by Sh̲āh Ṭahmāsp, he made his way towards Sultaniya, the first meeting with Ṭahmāsp taking place somewhere near that city around July 1544.[17] He stayed in that area for about two months,[18] his official host being the friendly Bahrām Mirzā, brother

14. Mirzā Ḥaidar's teacher is not the same person as a pupil of Bihzād with the same name, see BWG, p. 85. A Darwi̱sh Muḥammad is known to have been working for Humāyūn at Kabul and one wonders if he is the same painter. See *infra*, p. 13.

15. Perhaps the same as Maulānā Yūsuf, a painter serving Humāyūn. See *infra*, p. 17.

16. The Patna *Tārīkh̲-i Kh̲āndān-i Tīmūriya*, fol. 298, mentions Kh̲wāja ʿAbd al-Samad and Mīr Sayyid ʿAlī as teachers of Humāyūn. See Muqtadir, *Catalogue*, vol. 7, p. 45. Sh̲āh Ṭahmāsp, his brother Bahrām Mirzā, and his nephew Ibrāhīm Mirzā are known to have practiced painting. See *Calligraphers and Painters*, p. 3, and Arnold, *Painting in Islam*, p. 33.

17. Ray, *Humāyūn in Persia*, p. 26.

18. *Ibid.*, p. 50.

of Sh̲āh Ṭahmāsp, and himself an artist and patron of painting.[19] It was at Sultaniya that Humāyūn seems to have met some of the calligraphers and painters who were later to take up employment under him. Mīr Muṣawwir and his son Mīr Sayyid ʿAlī must have been among these, for Qāẓī Aḥmad,[20] in his treatise on calligraphers and painters, quotes him as saying to Sh̲āh Ṭahmāsp, "If that sultan of the universe gives me Mīr Muṣawwir, I shall send him from Hindustan one thousand *tumans* as a present."[21] In view of this favourable notice, our author continues, both proceeded to India, though Sayyid ʿAlī, the son, was the first to do so. Another artist who joined Humāyūn and went to India, according to Qāẓī Aḥmad, was Dost-i Dīwāna, though we do not know where he was recruited.[22] K̲h̲wāja ʿAbd al-Ṣamad also appears to have met Humāyūn at Sultaniya. The king is reported to have appreciated his paintings and apparently offered him an appointment.[23] Humāyūn then moved to Tabriz where he made a short halt of five days. The *Akbar-nāma* describes K̲h̲wāja ʿAbd al-Ṣamad as entering the imperial service there, though he was not able to immediately accompany Humāyūn due to "the hindrances of fate."[24] Tabriz was the furthest west Humāyūn was to travel. He then started back with Sistan as his destination, where troops promised by Sh̲āh Ṭahmāsp were to join him, stopping at Mashhad and other places en route. In March 1545, he laid siege to Kandahar held by Mirzā ʿAskarī, his brother, captured it in September, and went on to seize Kabul a couple of months later (15 November 1545) from yet another brother Mirzā Kāmrān. Here he was reunited with his infant son Akbar.

19. Jauhar, p. 65. Other accounts make out Bahrām Mirzā as being hostile to Humāyūn but we trust Jauhar's testimony as being that of an eyewitness. See Ray, *Humāyūn in Persia*, p. 35.
20. Qāẓī Aḥmad was the son of Sh̲araf al-Dīn Ḥusain Qūmī known as Mīr Mun̲sh̲ī, a servant of Ibrāhīm Mirzā (ca. 1543—70), the nephew of Sh̲āh Ṭahmāsp and Governor of Mashhad. He was a scholar rather than an artist, and wrote serveral works, among them an account of calligraphers and painters known in Persia as the *Gulistān-i Hunar*, "Rose Garden of Art," in 1596—7. In view of his close association with Ibrāhīm Mirzā who was a man of numerous talents and a great patron of the arts, his accounts of contemporary and sixteenth century personages is quite reliable. See Zakhoder's introduction to *Calligraphers and Painters*, pp. 1—27.
21. *Ibid.*, p. 185. According to Welch, *King's Book of Kings*, p. 71, Sh̲āh Ṭahmāsp's disaffection with painting began ca. 1544—45, just around the time Humāyūn arrived at the Persian court. This would account, to some extent, for the success Humāyūn had in attracting away some of the leading talent from the Safawi court.
    Gray, *Painting of India*, p. 77, misinterprets Qāẓī Aḥmad's report as an attempt by Humāyūn to bribe one of the Sh̲āh's leading court painters. If so, it would be indeed odd for Humāyūn to tell the Sh̲āh about it. All that Humāyūn seems to be saying to the Sh̲āh is that he would take good care of the painter in case he was permitted to take up employment with him.
22. *Ibid.*, p. 180.
23. Chaghatai, "K̲h̲wāja ʿAbd al-Samad," p. 158.
24. *Akbar Nāmā* I, pp. 444—45.

The Kabul interregnum

Kabul became the center of operations for ten years, until November 1554 when Humāyūn finally set out on the long awaited march towards Delhi. But the interim was hardly a peaceful interlude. There were futile expeditions against neighbouring rulers and he was much harried by Kāmrān who seized every opportunity to raise a revolt, actually managing to occupy Kabul a few times, though only for very short periods. Opportunities were provided by Humāyūn's ill-fated expeditions to Badakhshan (1547) and Balkh (1549) from where he had to retreat hastily due to one of Kāmrān's threatening moves against Kabul. It was after this retreat, sometime after September 1549, that he received at Kabul the painters Mīr Ṣayyid ʿAlī and Khwāja ʿAbd al-Ṣamad who arrived by way of Kandahar.[25] Shortly after (1550), occurred the battle of Kipchak between the forces of Humāyūn and Kāmrān in which Humāyūn was severely wounded. Khwāja ʿAbd al-Ṣamad was apparently present during the battle, and, in the ensuing confusion, got separated from the emperor, though he joined him shortly afterwards.[26] More serious was the loss of the royal books, but by good fortune they were recovered soon after when Kāmrān was defeated in the fight at Ushtar Gram (1550),[27] though this did not put an end to his mischief.

According to Bāyazīd who was with Humāyūn in these times,[28] quite a group of artists had gathered around the emperor at the Kabul court. In addition to Khwāja ʿAbd al-Ṣamad and Mīr Sayyid ʿAlī, are mentioned Mullā Dost,[29] whose skill Bāyazīd praises greatly, Maulānā Darwīsh Muḥammad, and Maulānā Yūsuf. Works *(ṣafās)* by all of them had been sent as gifts to ʿAbd al-Rashīd Khān, the ruler of Kashghar

25. *Ibid.,* p. 552. Also see Appendix A.
26. *Ibid.,* p. 558.
27. *Ibid.,* p. 571. Two unattended camels loaded with boxes being seen on the battlefield, Humāyūn claimed them as his personal share of the plunder. When they were opened, "by a beautiful coincidence it was found that the special, royal books which had been lost at the battle of Qibcāq (Kipchak) were in these boxes and in perfect condition. This was the occassion for a thousand rejoicings."
28. For an account of Bāyazīd and a translation of his important account of painting at the Kabul court, see Appendix A, pp. 171–73.
29. Chaghatai, "Khwajah ʿAbd al-Samad," pp. 158 ff., for some unexplained reason renders this name as Dost Muḥammad. We are particularly puzzled as he seems to have used the same text as us namely the one published by Hidayat Hosain. The name there is clearly Mullā Dost, no Dost Muḥammad being at all mentioned. Welch, *King's Book of Kings,* p. 172, suggests that a Dost Muḥammad who worked on the famous *Shāh-nāma* of Shāh Ṭahmāsp went to India where a later version of one of his paintings in the *Shāh-nāma* (521 verso, *The story of Haftwād and the worm*) was made for the emperor Jahāngīr. I do not know if Dost Muḥammad did go to India but Bāyazīd certainly does not say so.

apparently in Rajab A. H. 959/June-July 1552.[30] Art was thus patronised at Kabul in spite of the troubled circumstances, many of them of Mirzā Kāmrān's making. The endless strife between him and Humāyūn, however, came to an end in August 1553 when he was handed over by Sulṭān Adham the chief of Ghakkar, and blinded. The road to India was finally clear. Starting out from Kabul sometime in November, Humāyūn's troops entered Lahore on the 24th February 1555, and on the 23rd July 1555, he once again seated himself on the throne of Hindustan. Barely six months later, on the 20th January 1556, he accidentally fell down the stairs and died.

Humāyūn's painters and their works

Whether Humāyūn had a full-fledged atelier before his exile in Persia is unclear, but one can presume that he did, for this would be in keeping with the general practice among the Muslim rulers of the time. There is one clear reference, however, to the presence of a painter with Humāyūn during the period when his situation was extremely difficult. Jauhar recounts an incident at Amarkot (August-September 1542) where the defeated Humāyūn had paused in the course of his wanderings. A bird flew into Humāyūn's tent; it was caught and he himself clipped its feathers; a painter was sent for to take its likeness; and the bird was subsequently released.[31] The existence of a library, of which the painting establishment was usually an adjunct, however, is well attested. We have already referred to the theft of Hātifī's *Tīmūr-nāma* during the Gujarat campaign and to the somewhat dangerous practice of taking books to battle. They were nearly lost at Kipchak (1550); and Jauhar reports an earlier incident when anxious enquiries were made by Humāyūn about the safety of his library after the battle with Kāmrān at Talikan (1548) during the Badakhshan campaign.[32]

Unfortunately, even if Humāyūn did have painters working for him before the Persian exile, no examples have as yet been discovered. All that we have are a few

30. ʿAbd al-Ras͟hīd K͟hān, commonly referred to as Ras͟hīd K͟hān was the ruler of Kashghar from 1533–66. He was descended from Yūnus K͟hān, Bābar's maternal grandfather and actually received his name from Bābar. He was a great friend of Mirzā Ḥaidar but treated his relatives very cruelly on his accession to the throne. In spite of this Mirzā Haidar named his history in his honour, calling it the *Tārīkh-i Ras͟hīdī.* Ras͟hīd K͟hān was reputed to be an excellent painter and once painted a tree so skilfully that even masters were astounded (*Tarikh-i-Rashidi,* p. 139). Humāyūn seems to have kept in close touch with him from Kabul. The arrival of an embassy from him (1548), despatch of letters informing him of Kāmrān's misbehaviour (1549), and his final end (1554) are recorded in the *Akbar Nāma* I, pp. 542, 551, 609. See *Tarikh-i-Rashidi,* introduction, pp. 120 ff. from which this account has been largely abstracted.
31. Jauhar, p. 43.
32. *Ibid.,* p. 90.

paintings subsequent to the Persian sojourn, and none earlier than the Kabul period. There should be a greater body of work, but little has been identified, if at all it has survived.

We will summarise below the information we have about Humāyūn's painters and their works:

1. Maulānā Dost.

Bāyazīd singles him out for the most praise, comparing him to Mānī as a painter, and also calling him accomplished in calligraphy and gilding. It would seem that Maulānā Dost had first taken up service under Humāyūn, but then went to Mirzā Kāmrān, being annoyed because the emperor had prevented him from drinking wine. He apparently returned soon afterwards, for a drawing by him was included among the gifts sent from Kabul to Rashid Khān.

A *Portrait of Shāh Abū al-Maʿālī* recently sold at an auction is probably by him, though the inscription presents some difficulties.[33] The style is entirely Persian and shows no Mughal features.

2. Dost-i Dīwāna.

Qāẓī Aḥmad praises a painter, Dost-i Dīwāna by name, an incomparable pupil of Bihzād, "perfect in skill and ability. He spent some time in the service of the monarch equal in dignity to Jamshīd (Shāh Ṭahmāsp?) after which he went to India and made much progress there."[34]

It is possible that this Dost-i Dīwāna is the same person as Bāyazīd's Maulānā Dost, the word *dīwāna* not being a part of the name proper, but an appellation meaning "ecstatic," "eccentric." Being a fine artist and a pupil of Bihzād, he would qualify for the high praise bestowed on him by Qāẓī Aḥmad, and his running away to Kāmrān and returning thereafter perhaps indicates a certain eccentricity of character appropriate to his name. According to Stchoukine, no paintings by him are known.[35]

33. *Catalogue* of an interesting collection of Oriental miniatures, the property of Sir A. W. Robert Dent, Sotheby and Co., Tuesday 11 April 1972, Lot 18, Pl. facing p. 11. Professor C. M. Naim reads the inscription as follows: L. 1 *allāh akbar* L. 2 *jannat āshiyāni* L. 3 *īn ṣūrat shabīh-i shāh abū al-maʿālī-i kāshghar īst* L. 4 *ki dar khidmat-i ḥaẓrat . . .qurb-i tamām dāshta-and* L. 5 *ʿamal-i ustād dost muṣawwir.* The manner in which the name of the painter has been phrased suggests that what we have is an ascription, not a signature. Professor Naim suggests that the ascription was probably written in Humāyūn's time, but amended after his death, his name being erased after the word *ḥaẓrat* in the fourth line, the post-mortem title *jannat āshiyānī* being added at the top. The word *allāh akbar* was also added at this time which suggests that the alterations were made in the reign of Akbar.

34. *Calligraphers and Painters,* p. 180.

35. Stchoukine, *Manuscrits Safavis,* p. 33

3. Maulānā Darwīsh Muḥammad.

He is mentioned by Bāyazīd as one of the artists present at Humāyūn's court in Kabul, a painting by him also being sent to Rashīd Khan of Kashghar in A. D. 1552.

Two painters of this name are known; one, Maulānā Darwīsh Muḥammad, pupil of Shāh Muẓaffar, and the other, Dārwīsh Muḥammad Naqqāsh, a pupil of Bihzād, who supposedly did Bihzād's works for him.[36] It is possible that the painter under reference is the former who was also the teacher in painting of Mirzā Muḥammad Ḥaidar.[37] The close relationship between Humāyūn and Mirzā Muḥammad Haidar, a first cousin of his father, has been indicated earlier, and one can speculate, with some reason, that it was through this association that he may have joined Humāyūn's staff. The suggestion gains credence in as much as Darwīsh Muḥammad seems to have been with Humāyūn sometime around 1548–9 when Mirzā Ḥaidar had resumed his loyalty to Humāyūn with the dispatch of the embassy of Mīr Samandar to Kabul.[38]

No paintings ascribed to him appear to be known.

4. Maulānā Yūsuf.

He is also mentioned by Bāyazīd as one of Humāyūn's artists at Kabul, whose work was sent as a gift to Rashīd Khan of Kashghar. Mirzā Ḥaidar refers to a Mullā Yūsuf as a pupil of Bihzād; he was a rapid worker accomplishing in ten days what others took a month to do. Once again the association with Mirzā Ḥaidar makes one suspect that this Maulānā Yūsuf may be the same artist who was described by him.[39]

No works ascribed to Maulānā Yūsuf are known.

5. Mīr Muṣawwir.

According to Dost Muḥammad, he was a Sayyid, and together with Āqā Mīrak, "painted in the royal library (of Shāh Ṭahmāsp) illustrating especially a royal *Shāh-nāma* and a *Khamsa* of Niẓāmī so beautiful that the pen is inadequate to describe its merits." Mīrak and he also adorned an arched *jām-khāna* (mirror house) for Prince Bahrām in the most exquisite and masterly manner, making it as beautiful as paradise peopled with fair youths and houris.[40] Qāẓī Aḥmad says he was a native

36. BWG, p. 191.
37. *Ibid.* See *supra,* p. 12.
38. Erskine, p. 366.
39. BWG, p. 191. See *supra,* p. 12. He is to be distinguished from Maulānā Yūsuf Ghulām-khāṣṣa, who according to Iskandar Munshī was appointed as librarian by Shāh Ṭahmāsp towards the end of his reign. See Arnold, *Painting in Islam,* p. 161.
40. BWG, p. 186. Dost Muḥammad, besides being a painter, was also the author of a short account of painters appended to an album made for Bahrām Mirzā, now in the Topkapu Serai Library. See BWG, p. 139, pp. 183–88 for an abstract.

of Badakhshan, "a portraitist working neatly, who made very pleasant and pretty images."[41] He was much admired by Humāyūn who tried to secure his services from Shāh Ṭahmāsp, as noted earlier; but it was his son Mīr Sayyid ʿAlī, who, according to the same author, first went to India. This seems to be confirmed by Bāyazīd who does not mention him among the artists present at Kabul. That he did reach India is attested by his portrait, ascribed to Mīr Sayyid ʿAlī, in the Musée Guimet, Paris, where he is shown holding a petition addressed to Akbar.[42]

The Turkish historian ʿĀlī says that Mīr Muṣawwir succeeded Sulṭān Muḥammad as chief of Ṭahmāsp's atelier, and that he was a pupil of Bihzād and teacher of Zain al-ʿĀbidīn. This is not confirmed by Persian historians, and Stchoukine has demonstrated the necessity of taking ʿĀlī's statements with much caution. Some authors have identified him with one Mīr Naqqāsh, but Stchoukine casts doubts on this also.[43]

Though a great and famous artist, very few pictures are known to be by him, or are reasonably ascribed to him. One of these could be the famous *Naushīrwān and the owls* in the British Museum *Khamsa* of Niẓāmī painted for Shāh Ṭahmāsp between A. H. 946–9/1539–43.[44] Three more paintings by him in the Ṭahmāsp *Shāh-nāma,* one of them dated A. H. 934/1527–8, have been recently published by Stuart Cary Welch who sees in him a painter with an immaculately brilliant palette, and graceful, hard-edged forms, adept in romantic themes and a specialist in portraying pretty girls and handsome youths.[45] Robinson also thinks that the *Portrait of Murkhān Beg Sarfarchi* in the British Museum (1930–11–12–02) is his work.[46] He was a conservative painter, and we do not know if he contributed anything to Mughal painting (see, however, Appendix E).

6. Mīr Sayyid ʿAlī.

According to Qāẓī Aḥmad, he was the son of Mīr Muṣawwir, and possessed even greater talents. Learning of the high regard Humāyūn had for his father, he was the

41. *Calligraphers and Painters,* p. 185.
42. Stchoukine, *Miniatures Indiennes,* pp. 11–12, Pl. IIa. It is not known when Mir Muṣawwir joined Mughal service.
43. Stchoukine, *Manuscrits Safavis,* pp. 32, 38–39.
44. The miniature has been reproduced in colour, Binyon, *Poems of Nizami,* Pl. III. On the wall of the ruined house is written a verse at the end of which is a line which is read by Arnold and Grohmann, *Islamic Book,* p. 80 as "inscribed by Mīra(k) the painter 946;" Sakisian, "School of Bizhad," p. 82 disagrees and reads the name as Mīr Muṣawwir. Stchoukine, *Manuscrits Safavis,* pp. 71, 191, agrees with Sakisian and accepts the painting as a work of Mīr Muṣawwir.
45. Welch, *King's Book of Kings,* pp. 100, 128, 168, 20. Of these fol. 516 verso is inscribed with his name and a date of A. H. 934/1527–28. The other two miniatures, fols. 28 verso, 67 verso, are attributions by Welch.
46. Robinson, *Persian Miniature Painting,* p. 38. For a reproduction see Chaghatai, "Mir Sayyid Ali," p. 27.

first to proceed to India.[47] Mīr ʿAlā al-Daula, the author of the *Nafāʾis al-Maʾāsir*, says that he was from Tirmiz, and though his ancestors had lived for some time in Badakhshan, he himself had spent his youth in Iraq.[48] He, together with his father, was a member of the atelier of Shāh Ṭahmāsp at least until 1543, working on the royal *Shāh-nāma* and also the British Museum *Khamsa* of Niẓāmī which was completed that year. It is not clear where he first met Humāyūn; but Bāyazīd informs us that a *farmān* was issued to him from Astalaf near Kabul; and we know that he presented himself to Humāyūn at Kabul in 1549, his work being included in the gifts sent to Rashīd Khān of Kashghar in 1552.[49] Both he and ʿAbd al-Ṣamad taught painting to Humāyūn,[50] and there is evidence to indicate that he was much preferred by him, and perhaps given the title of *nādir al-mulk.*[51] Mīr Sayyid ʿAlī accompanied Humāyūn to India, and continued to work under Akbar, being in charge of the great project of illustrating the *Qiṣṣa-i Amīr Ḥamza* for about seven years, and completing about four volumes.[52] His reputation was great, Abū al-Faẓl considering him as one of the forerunners on the high road of knowledge;[53] and Badāyūnī calls him a second Mānī in India, each page of whose paintings is a masterpiece.[54] He seems to have left for a pilgrimage to Mecca shortly after A. H. 979/1571–2.[55] His future career is unknown, though according to one source, he died there.[56]

Mīr Sayyid ʿAlī also wrote poetry under the pen-name of *Judāʾī.* According to Ṣādiqī Beg Afshār, a Persian painter and author of the late sixteenth century, he and Maulānā Ghazzālī of Mashhad composed satirical verses against each other, but were finally reconciled on Mīr Sayyid ʿAlī's making a portrait of the Maulānā.[57] It was also alleged by some that he misappropriated the verses of one Ashkī, who gave them to him on his death bed, but this could be mere gossip.[58]

47. *Calligraphers and Painters,* p. 85.
48. See Appendix B, fn. 12. Abū al-Faẓl refers to him as being from Tabriz which may be an error or a reference to his having made that city his home when employed under Shāh Ṭahmāsp.
49. Appendix A. The Mīr's arrival at this time is confirmed by the *Akbar Nāmā* II, p. 525.
50. See fn. 16, *supra.*
51. The evidence for this is scanty, the information occurring only as an interpolation in the *Nafāʾis al-Maʾāsir,* see Appendix B, fn. 16. The title also appears in the Binney *Portrait of a young Indian scholar* where it may perhaps be a later addition. Abū al-Faẓl does not indicate that he possessed this title.
52. See Appendix B, p. 180.
53. Appendix C, p. 183.
54. Badāyūnī III, p. 292.
55. Appendix B, p. 178.
56. Chaghatai, "Mir Sayyid Ali," p. 27 quoting *Tuhfa al-Kiram,* Bombay 1886–7, II, p. 176.
57. *Tazkira Majmaʿ al-Khwāṣ* of Ṣādiqī Beg Afshār, Tabriz 1909, p. 97, quoted in Chaghatai, "Mir Sayyid Ali," p. 26. For a short account of the extraordinary personality of this artist see Arnold, *Painting in Islam,* p. 142.
58. *Āʾīn* I, p. 667, fn. 4.

Robinson assigns to him the *Prince and a page* in the British Museum (1930–11–12–01) produced at an earlier stage of his career, about 1535, the painting being derived from the work of his father.[59] Recent researches by Welch on the *Sh̲āh-nāma* of Sh̲āh Ṭahmāsp are beginning to throw considerable light on the formative phases of his style when he was beginning to emerge as a skilled master in the royal Persian atelier. His work there is characterised as that of a brilliant designer, possessed of great technical virtuosity and accuracy of observation.[60] His mature Persian phase is represented by four miniatures, all of which appear to have belonged to a copy of the *Kh̲amsa* of Niẓāmī prepared for Sh̲āh Ṭahmāsp between October 1539 and April 1543.[61] In all of these expressive works, full of well observed life of human beings, we find a trait that in a much more developed form becomes a constant feature of early Mughal art.

Of the artist's paintings of the Mughal phase, which concern us the most, we know hardly anything. Only two miniatures can be tentatively attributed to him with any plausibility. One of these is the *Portrait of a young Indian scholar* now in the Binney collection (Pl. 64).[62] Here we see a young man wearing a Deccani-type turban and earrings, seated in an open landscape, bending over a bookstand. On a writing tablet lying by his side is inscribed a verse, "On every tablet, it is written in gold, the reprimand of the teacher is better than the love of the father." Below, in small but elegant letters is written, *ṣūrat sayyid ᶜalī nādir al-mulk humāyūn sh̲āhī.*

The painting is of importance, as the style of the work marks a new phase distinct from Persian painting, showing in the more vigorous movement of line, feeling for spatial depth, voluminousness, and boldness of execution, characteristics associated with the *Ḥamza-nāma* to which period this miniature would seem to belong, a date around 1560–65 being most likely. Further affinities to the *Ḥamza-nāma* can be seen in the softly modelled rocks in the background and the fluffy texture of the scarf wound around the shoulders.

If the painting is of the date we think it is, some difficulties are presented by the inscription which gives the appearance of being a signature written at the time of

59. Robinson, *Persian Miniature Painting,* p. 54, Pl. 17.

60. Welch, *King's Book of Kings,* pp. 136, 179. One entire painting, fol. 568r, *Bahrām Gūr pins the coupling onagers* (p. 176) has been attributed to him by Welch, and he also sees his hand in fol. 37 verso, *The death of Ẓaḥḥāk,* and fol. 102 verso, *Qāran slays Barman* (p. 136).

61. Of these only one is in the manuscript itself, *Majnūn brought before the tent of Lailā* (fol. 157r), reproduced in colour, Pl. XII of Binyon, *Poems of Nizami.* Two miniatures, previously identified as scenes of urban and nomadic life, but in reality representing *Alexander giving an entertainment* and *A conference between the tribes of Naufal and Lailā* are in the Fogg Art Museum (*Survey of Persian Art,* Pl. 908 A and B; 909 A and B) while the fourth, *Kh̲usrau Parwīz battling Bahrām Ch̲ūbīna* is in the Royal Scottish Museum, Edinburgh (Robinson, *Persian Miniature Painting,* p. 40). See Welch, *King's Book of Kings,* p. 64.

62. See Binney, *Catalogue Mughal and Deccani Schools,* pp. 26, 30; Welch, *Meadow,* pp. 88–89.

Humāyūn *(humāyūn shāhī),* omitting as it does the honorifics such as *ustād* or *mīr* or both, which are usual in ascriptions. The painting would then have to be considered earlier, belonging to the period of Humāyūn. Theoretically, it could have been painted at any time between 1549, the date of the artist's joining Humayun at Kabul, and the 24th of January 1556, the day of Humāyūn's death, a period nearer to 1556 being preferable in view of the striking non-Persian features. On the other hand, in spite of the inscription, the painting could well be of the early Akbar period as we think it is, the words *humāyūn shāhi* not necessarily indicating that it was done in the reign of Humāyūn. The inscription, whether a signature or an ascription, could have also been added on when the miniature was prepared for mounting in an album, as happened, for example, in the case of some of Khwāja ᶜAbd al-Ṣamad's miniatures when they were remounted for inclusion in the *Muraqqaᶜ-i Gulshan.*[63]

One other work possibly by Mīr Sayyid ᶜAlī is the *Portrait of Mīr Muṣawwir* in the Musée Guimet, Paris.[64] An old man wearing a pince-nez bends over a long scroll of paper which he holds in his hands. On it is inscribed a petition in minute but legible handwriting.[65] The petitioner is there stated to be Mīr Muṣawwir, who stresses his age and prays for favours on his son who has been in service for a long time. He himself, God willing, hopes to hasten to the royal presence.

The picture has an ascription on the margin which reads *ᶜamal mīr sayyid ᶜalī,* the work of Mīr Sayyid ᶜAlī.[66] It shows many similarities of design to the *Portrait of a young Indian scholar* in the Binney collection (Pl. 64), but the work is much more delicate, lacking the boldness of the *Portrait*. It is difficult to judge the date of the painting, the petition giving no indication if it was being addressed to Humāyūn or Akbar. If it is indeed by Mīr Sayyid ᶜAlī, it is likely to be of about the same period as that artist's *Young Indian scholar,* or a little earlier, being more Persian in execution.[67]

63. See *infra,* p. 25 and p. 24, fn. 86. For later signatures of the same format as this one, cf. the one by Manṣūr in a miniature forming part of a royal album in the Metropolitan Museum of Art (55.121.10.12): *ᶜamal banda-i dargāh manṣūr nādir al-ᶜaṣr; jahāngīr-shāhī.*
64. Stchoukine, *Miniatures Indiennes,* pp. 11–12, Pl. IIa.
65. For a reading and French translation of the inscription by Minorsky see *ibid.,* p. 12.
66. Stchoukine, *Miniatures Indiennes,* p. 12.
67. It is also possible that the ascription is incorrect and that the painting was done by Mīr Muṣawwir himself, this being suggested by the inappropriateness of Mīr Sayyid ᶜAlī painting a portrait of his father presenting a petition in which it is he (Sayyid ᶜAlī) who is the beneficiary, this being contrary to etiquette. If Mir Muṣawwir wished to present a petition in person, and he was unable for some reason to do so, the appropriate thing for him to do would be to send a self-portrait holding the petition by way of proxy rather than have his son, on whose behalf he was pleading, make the portrait for presentation to the king. Mīr Muṣawwir's skill in portraiture has been referred to by Qāẓī Aḥmad, and if this portrait is indeed by him, it is still evident in India when he must have been a fairly old man.

7. Khwāja ʿAbd al-Ṣamad.

Of the painters who began their Indian careers with Humāyūn, ʿAbd al-Ṣamad emerges the most clearly, partly because his career at the Mughal court was a long one, and partly because he served as an officer of some rank in capacities other than those of a painter. His son Muḥammad Sharīf, himself an accomplished painter,[68] was a great favourite of the future emperor Jahāngīr, being appointed by him, no sooner than he came to the throne, to the highest available rank, that of Amīr al-Umarāʾ. Yet annother son named Bihzād was also a painter, though not of great merit.[69]

Khwāja ʿAbd al-Ṣamad was of noble descent, son of Khwāja Niẓām al-Mulk, vizier to Shāh Shujāʿ, a governor of Shiraz.[70] With the possible exception of a painting in the *Shāh-nāma* of Shāh Ṭahmāsp,[71] there is little evidence to indicate that he was a member of Shāh Ṭahmāsp's atelier. He had the title of *shīrīn-qalam* (sweet-pen/brush), which, according to Jahāngīr, was conferred upon him by Humāyūn.[72]

He entered service at Tabriz, but for some reason was unable to join Humāyūn immediately. This he did at Kabul in 1549 together with Mīr Sayyid ʿAlī. He was, as we have said earlier, with Humāyūn at the Battle of Kipchak (1550), got separated from the emperor in the confusion, but was able to rejoin him shortly afterwards. According to Bāyazīd, his handiwork was also among the gifts sent out to Rashīd Khān in 1552;[73] and according to the *Tārīkh-i Khāndān-i Tīmūriya,* he, together with Mīr Sayyid ʿAlī, taught painting to Humāyūn. That Akbar also learned painting from him at Kabul and Delhi is testified to indirectly by two miniatures in the *Muraqqaʿ-i Gulshan* and also by the *Akbar-nāma.*[74] He followed Humāyūn to India and continued to serve under Akbar,taking charge of the *Ḥamza-nāma* project after Mīr Sayyid ʿAlī had been at work on it for seven years and had completed at least four volumes. This occurred shortly after A. H. 979/May 1571. He is known to have been close to Akbar; and Jahāngīr recalls his intimacy with the emperor and the great

68. Three paintings designed by him occur in the Jaipur *Razm-nāma,* see Hendley, Pls. 57, 101, 104.
69. A painting formerly in the Kevorkian Foundation, sold in 1967 (*Catalogue* of highly important Oriental manuscripts and miniatures, Sotheby and Company, 6 December 1967, Lot 111) has an ascription identifying him as the son of ʿAbd al-Ṣamad. The miniature is very much in the style of the father. Another painting by Bihzād, corrected by his father, is in the British Museum *Dārāb-nāma* (fol. 14v, Pl. 45).
70. Mahfuz-ul Haq, p. 241.
71. Welch, *King's Book of Kings,* pp. 184 ff. attributes fol. 742v, *Assasination of Khusrau Parwīz* to ʿAbd al-Ṣamad.
72. *Tūzuk* I, p. 15.
73. Appendix A, p. 173. BWG, No. 232, Pl. CV-B is a painting dated 1551 and thus of his Kabul period.
74. BWG, No. 230, Pl. CIV-B and No. 232, Pl. CV-B. Both miniatures are by ʿAbd al-Ṣamad. In one Akbar is shown painting and in the other he is presenting a miniature to his father. Also see *Akbar Nāmā* II, p. 67.

dignity he had achieved in his council.[75] Abū al-Faẓl states that he became a grandee, attaining to the rank of 400 horse,[76] and praises his perfection as being due to the "transmuting glance of the king [which] has raised him to a more sublime level," a statement that hints at the positive part played by Akbar in moulding the style of his painters and ʿAbd al-Ṣamad's receptivity to the tastes of his patron. The Khwāja is also singled out as a good teacher, his pupils becoming masters under his tutelage, the most famous being the painter Dasavanta.[77] A curious incident reported in the *Maʾās̱ir al-Umarāʾ* also shows that he may have done some wall paintings for the Khan-i Aʿẕam, Mirzā ʿAzīz Kokā.[78] In the twenty-second year of the reign (1577–8), he was put in charge of the mint reforms instituted by Akbar, at the same time being made master of the mint of the capital city of Fatehpur-Sikri, the importance of this appointment to be gauged by the fact that the provincial mints were put in charge of such great officers as Muẕaffar Khān, Rājā Ṭodaramala, and Khwāja Shāh Manṣūr.[79] In the twenty-seventh year (1582–3), he was entrusted with the task of the buying and selling of leathern articles;[80] and in the twenty-eighth year (1583–4), he was appointed to assist Sulṭān Murād, the emperor's son, in the management of the royal household.[81] In the thirty-first year (1586–7), he was appointed to the provinces, being elevated to the high position of *dīwān* of Multan.[82] This is the last official notice we have of him. Bāyazīd reports having met him in Lahore in A. H. 999/1590–1;[83] and Fazūnī of Astarabad, a traveller to India, declares that Khwāja ʿAbd al-Ṣamad made and presented to Akbar a marvellous mirror box, somewhat like a kaleidoscope, in A. H. 1008/1599–1600.[84] It is interesting to note that in spite of his other duties, he never ceased to paint and continued to contribute to the productions of the royal atelier, the last known painting by him being present in the *Khamsa* of Niẕāmī dated 1595.

Khwāja ʿAbd al-Ṣamad seems to have been a pious Shīʿa, a member of the Dīn-i Ilāhī, and, like others close to the emperor, of liberal religious views. The bigoted Badāyūni describes him as being much "occupied with ceremonial prayers and fasts, and supererogatory prayers and outward devotions." He had faith in one Ḥājī

75. See Appendix C, p. 183. *Tūzuk* I, p. 15.
76. *Āʾīn* I, p. 554.
77. Appendix C, p. 183.
78. *Maāthir* II, p. 818. The incident is based on the *Zakhīra al-Khwānīn* of Shaikh Farīd of Bhakkhar, a work written in 1650 and not considered to be very reliable by the author of the *Maāthir, ibid. I,* p. 7.
79. *Akbar Nāma* III, p. 321.
80. *Ibid.,* p. 585.
81. *Ibid.,* p. 598.
82. *Ibid.,* p. 779.
83. Appendix A, p. 172.
84. Quoted in Chaghatai, "Khwajah ʿAbd al-Samad," p. 177.

Ibrāhīm, a teacher of the traditions of the prophet, though he was exhorted by him to have love in his heart for the orthodox successors of the Prophet, without which "all these observances will profit you nothing."[85]

Several paintings, some of them signed, some ascribed and yet others attributable to him, are known and published. These belong both to the Persian phase of his career and to the period of Humāyūn and Akbar, so that it is possible to get a firmer idea of his style than of any other painter who served Humāyūn.

Of Khwāja ʿAbd al-Ṣamad's paintings produced in Persia, even fewer are known than those of Mīr Sayyid ʿAlī. Besides the miniature in the Ṭahmāsp *Shāh-nāma* attributed to him by Welch, there is a painting in the *Muraqqaʿ-i Gulshan* depicting a *darwīsh* praising God while his companions sleep that bears his signature.[86] Yet another miniature in the same album, depicting princes in a pavilion, has been tentatively attributed to him by Stchoukine on grounds of style.[87] All of these distinguish him as an artist of considerable merit, with a fine sense of design, meticulous execution and comparatively realistic observation, somewhat like Mīr Sayyid ʿAlī, but a little drier, and lacking his poetry.

Khwāja ʿAbd al-Ṣamad's style does not seem to have changed much during the period he served Humāyūn. One of the two paintings which are of this time shows two boys (the one who paints being undoubtedly Akbar) in a landscape, and bears an inscription which declares it to have been painted in just half a day on Nauroz A. H. 958/9 January 1551 by Maulānā ʿAbd al-Ṣamad.[88] The honorific *maulānā* is somewhat unusual for him, but the painter seems also to have been so addressed, for example, by Bāyazīd.[89] The work is sketchy and lightly coloured and has indeed all the appearance of being quickly finished. Besides being the earliest dated work of the Khwāja of which we are aware, it is also the first painting known to have been done by him for Humāyūn. The style is still very Persian, and the provenance is Kabul where Humāyūn was in residence at the time.

The other picture of this period is the splendid *Akbar presenting a miniature to*

85. *Badāyūnī* III, p. 196.

86. The miniature, which is included in the Muraqqaʿ-i Gulshan now in the Imperial Library, Teheran, is reproduced in BWG, No. 231, Pl CV-A. A decorative rhyming label in the bottom left corner translates, "the slave who writes a broken hand *(shikasta-raqam),* ʿAbd al-Ṣamad *shīrīn-qalam.*" Stchoukine, *Manuscrits Safavis,* p. 85, thinks it to be Tabriz, mid-sixteenth century, while Ettinghausen, "ʿAbdu'ṣ-Ṣamad," p. 18 relates it to a manuscript of the *Haft Aurang* of Jāmī dated 1556–85 in the Freer Gallery of Art. The label, which appears to be a signature, would not contradict either of these dates, for it appears that some of ʿAbd al-Ṣamad's signatures to his own works in the *Muraqqaʿ-i Gulshan* were added at the time the *Muraqqaʿ* was compiled.

87. Stchoukine, *Manuscrits Safavis,* p. 78, assigns it to Tabriz, ca. 1540. For a reproduction see BWG, Catalogue No. 165, Pl. XCIV-B.

88 BWG, No. 232, Pl. CV-B.

89. Appendix A, p. 173.

Humāyūn,[90] still basically Persian, but with a strong connected rhythm flowing throughout, fairly brisk movement and a crowded composition which may perhaps indicate a slight movement away from Safawi origins. Judging from the age of Akbar, who appears to be about twelve years old, and also by the presence of Humāyūn, the picture should be dated to ca. 1554, probably just before the departure for India, which is quite consistent with the style. An inscription, written in very small characters on a little book placed on the tiled floor, reading *allāh akbar al ʿabd ʿabd al-ṣamad shīrīn qalam,* is a little puzzling as it is phrased like a signature, and would seem to indicate that the miniature was made in the reign of Akbar. This, however, is belied by the style, and, in all probability, the inscription was added when the miniature was mounted for the *Muraqqaʿ-i Gulshan* when ʿAbd al-Ṣamad was still alive.[91] The verses at the top and bottom were also written at this time stating that the miniature was drawn by the brush of ʿAbd al-Ṣamad, and that the picture which was being presented by Akbar to his father repeated on a minute scale the scene represented in the larger painting itself.

It is interesting to observe that even in miniatures done during the reign of Akbar, the Khwāja retains strong Persian traces, especially the decorative emphasis, to a very late period, never quite achieving the full blown movement, plasticity, and psychological insight so typical of the Mughal style. He prefers to draw small and dainty figures and shows a pronounced preference for rich, rocky outcrops, skilfully built up in small units,[92] all in rich and delicate colours, everything unfailingly exquisite and elegant (Pl. 65). His earliest painting from the reign of Akbar is dated 1557, but unfortunately no photograph has been published.[93] The others, such as *A horse led by a groom,*[94] *A prince hunting with falcons,*[95] *Jamshīd writing on a rock* (dated 1588) (Pl. 65),[96] *A royal hunt,*[97] and *Khusrau hunting* (dated ca. 1595),[98]

90. BWG, No. 230, Pl. CIV-B. A colour reproduction is to be found in *Survey of Persian Art* V, Pl. 912.

91. The phrase *allāh akbar,* a very ancient sacred formula of Islam, seems to have been given a special currency from about 1579 onwards when it was used in the first *khuṭba* read by Akbar in the great mosque at Fatehpur-Sikri. The compilation of the *Muraqqaʿ-i Gulshan* seems to have begun at least as early as Ramaẓān 1008/1600 which is the earliest date discovered on the borders. See Godard, "Les marges," p. 13.

92. It is interesting to observe that his son Bihzād, both in the Sotheby miniature (*Catalogue* 6 December 1967, Lot 111), and the *Dārāb-nāma* miniature (Pl. 45) paints a similar landscape.

93. BWG, No. 233.

94. Ettinghausen, "ʿAbduʾṣ-Ṣamad," p. 15. The painting bears an inscription, *ʿabd al-ṣamad shīrīn-qalam.*

95. Welch, *Meadow,* Pl. 15. The painting bears a late ascription to Mīr Sayyid ʿAlī but Welch correctly attributes it to ʿAbd al-Ṣamad.

96. The painting once formed part of the *Muraqqaʿ-i Gulshan* and is now in the Freer Gallery of Art. It bears a signature, *ṣūrat ʿabd al-samad shīrīn-qalam* and is dated both in the Hijri era (996) and the regnal year (32). A note states that it was done "in his old age."

Notes 97 and 98: see following page.

are all elegant but dry, and demonstrate a continuing attachment to Persian mannerisms which one would hardly expect in a painter who supervised the production of almost three-fourths of the *Ḥamza-nāma.*[99]

Princes of the House of Tīmūr

Having considered some of the painters who are known to have been in the service of Humāyūn, we can now turn to a very famous painting, the so-called *Princes of the house of Tīmūr* in the British Museum (No. 1913–2. 8. 1, Mts. 45 ins. by 42 ins.).[100] This unusually large work, painted on cloth, has strong Persian features; and as Mīr Sayyid ʿAlī and Khwajā ʿAbd al-Ṣamad were the two Persian painters known to have been working for Humāyūn, it was assigned by scholars to either one or the other of them. It seems to us, however, that of the two, the painting has much greater resemblances to the work of Khwāja ʿAbd al-Ṣamad, as exemplified in *Akbar presenting a miniature to Humāyūn,* than it does to Mīr Sayyid ʿAlī's *Portrait of a young Indian scholar,* and is therefore more likely to be by him. The painting also has affiliations with later work by ʿAbd al-Ṣamad, particularly in the treatment of the trees, the rich colour, and the calm postures of the figures, all of which further strengthen an attribution to him.

In the centre of the painting is a hexagonal throne with Humāyūn to the right facing the seated emperors Akbar, Jahāngīr and Shāh Jahān. On the balustraded pathway leading to the throne are pages and footmen carrying food and drink. Behind is a garden dominated by a large chinar tree, though in it also grow cypresses and flowering bushes and plants. To the left, are a group of cooks; to the right, is a gardener and an attendant. The pink mountains are painted against a golden sky with flying herons and other birds. In the foreground, which is severly damaged, are two diagonal rows of seated figures.

97. This miniature is also from the *Muraqqaʿ-i Gulshan* and is now in the Los Angeles County Museum. See *Heeramaneck Collection,* No. 198. Beach, *ibid.,* p. 143 attributes it, probably correctly, to ʿAbd al-Ṣamad. He has not noticed an ascription giving the name of the painter and other information beginning *ʿamal-i murīd* . . . which we have not been able to decipher.

98. Reproduced in Martin, Pl. 179, and Ettinghausen, "ʿAbduʾṣ-Ṣamad," Pl. 17. The painting is from a Khamsa of Niẓāmī dated 1595 now in the British Museum. It has an inscription on the margin, *ʿamal khwāja ʿabd al-ṣamad.*

99. One painting by ʿAbd al-Ṣamad is in the Jaipur *Razm-nāma* but is unpublished. The *Arrest of Shāh Abū al-Maʿālī* (Bodleian Library, Oxford, Ouseley Add. 172, fol. 4, reproduced in Ettinghausen, "ʿAbduʾṣ-Ṣamad," Pl. 16) is incorrectly ascribed to him. A Persian painting of a horse and a groom has been ascribed to ʿAbd al-Ṣamad (Sakisian, Fig. 55) but Stchoukine advises caution in accepting its correctness, *Miniatures Safavis,* p. 84.

100. For two reproductions, one of them in colour, see Binyon, *House of Timur.* He also briefly discusses the painting and considers it to be a Persian work painted at Kabul about 1550 either by Mīr Sayyid ʿAlī or Khwāja ʿAbd al-Ṣamad.

Several of the figures, though not all of them, have ascriptions. Those on the left are identified as Mīrān Shāh, Sulṭān Muḥammad, Abū Saʿīd, ʿUmar Shaikh, Bābar, Humāyūn, and Mirzā Kāmrān; those on the right as Abū Bakr, Bāʾisunghur, and Shāh Rukh. They are all Timurids, and it is probably this feature that accounts for the identification of the painting by scholars as a representation of the princes of the house of Tīmūr.

Unfortunately, these labels are entirely imaginative and cannot be accepted as reliable. The personages, if the labels are correct, are seated more or less in hierarchical order on the left from father to son, though the presence of a person identified as Mirzā Kāmrān is hard to explain, as is the repetition of Humāyūn, whose portrait, as well as that of his father (identified according to the labels), bears no resemblance to known portraits of the two. The labels on the right, on the other hand, demonstrate a very confusing seating order. We first have Abū Bakr, a son of Mīran Shāh, followed by Bāʾisunghur and Shāh Rukh. No lineal descent is to be seen, Bāʾisunghur being the son and not the father of Shāh Rukh, Shāh Rukh himself being a son of Tīmūr and not of Abū Bakr who was a nephew. All sense of propriety and decorum would also be violated if the ancestors were indeed sitting on the floor in subordinate positions, while Humāyūn occupied a position of exalted eminence on the throne. In fact, there is little to indicate, except for the labels, that the painting originally represented anything but a grand entertainment at the court of Humāyūn, showing the emperor, probably with his son Akbar seated in the place of honour in front of him,[101] the other figures representing the nobles of Humāyūn. The scale and magnificence of the painting suggests that it may have been painted to commemorate some great event, such as, perhaps, the thanksgiving entertainments on Humāyūn's return to the throne of Hindustan.

The picture originally seems to have belonged to the last years of Humāyūn, but apparently went through several stages of updating and revision. Most of the landscape setting and figures are original, though only the faces of Humāyūn on the throne, the seated courtiers, and the group of three pages standing after the first two figures immediately behind Humāyūn (Pl. 66) are contemporary with the original painting. The workmanship of these heads is distinctively Persian, especially of the three pages who are almost identical to those in *Akbar presenting a miniature to Humāyūn,* and may be contrasted with the rendering of the attendants who stand immediately behind them. Their heads are emphatically modelled, the lips full and the features clearly and delicately drawn, the differences to be understood as the

101. Stchoukine, *Peinture Indienne,* p. 116, thinks that here may have once stood the portrait of Tīmūr but this is hardly possible for he would then be in a subsidiary position to Humāyūn. The only place for Tīmūr would be the most honoured one, namely one where Humāyūn is shown seated on the throne.

result of retouching done later in the reign of Akbar (Pl. 66). Early in the reign of Jahāngīr, at a date not too distant from the Victoria and Albert *Ajmer darbār* and the Musée Guimet *Jahāngīr holding the portrait of his father,*[102] were added the figures of Akbar and Jahāngīr, replacing probably a portrait of the child Akbar. Jahāngīr may have wished to be included in the scene, and, feeling the incongruity inherent in being placed next to a father who was much younger than him, also had Akbar's portrait redone showing the appropriate relationship of age between father and son. The portrait of a prince who appears to be Khusrau, standing immediately behind Humāyūn, and that of Parwīz, standing behind Jahāngīr, were also added about this time; and some more heads were repainted, like that of the servant receiving food from the cook to the extreme left. Finally, the portrait of Shah Jahān was added, probably around the time of his accession to the throne.

In spite of these changes, the picture retains much of its original character for the main figures seem to be largely unretouched, and the colour has all the brilliance of a painting like *Akbar presenting a miniature to Humāyūn* by Khwāja ʿAbd al-Ṣamad, to which, as we have remarked earlier, it is closely related. An attribution to this artist is strengthened by the great chinar tree, its leaves shimmering against a golden sky, a Persian device much favoured by the Khwāja. Not only is it found in the earlier *Presentation scene,* but even closer parallels, right down to the nest ensconced between the branches, occur in his later pictures, like the Freer *Jamshīd writing on a rock* (Pl. 65) and the Los Angeles *A royal hunt (Heeramaneck Coll.*, p. 143). The painting, though basically Persian, seems to have been done after the *Presentation* in view of the more plastic rendering of the trunks of trees, the foliage and the flowers. The large size and the cloth carrier point in the direction of the *Ḥamza-nāma,* but the style is essentially in a much more conservative vein, and makes little attempt to strive for that vitality and power which the *Ḥamza-nāma* so successfully attained.

Humāyūn and the origins of the Mughal style

The account of Humāyūn's life, the cultural traditions he was heir to, the intellectual and artistic milieu in which he functioned, and what meager information we have been able to scrape up about his painters and their works seem to suggest a strong inclination and preference for the Persian style of work. The artists about whom we have any information were all imported from abroad, and even if Humāyūn had Indian artists in his employ, it is difficult to conceive of them as any other than those working in the style of the Indo-Persian group of manuscripts with

102. Stchoukine, *Peinture Indienne,* Pl. XXVIII and *Miniatures Indiennes,* Pl. VI.

its strong Persian features. The overall impression we get is that of a small and compact atelier consisting of a few artists who formed a somewhat closed group, working in an elitist and fairly rarefied atmosphere. The foreign artists, in particular, were strangers to India and Indian traditions, and even if they had the inclination, they had hardly the time to grow roots in Indian soil or to allow the environment to affect them, though they may have been able to accommodate themselves to the tastes of their patron in the six years or so they were with him. Their style, thus, for the most part, is a full blown Persian style bodily transplanted to India. It is often quite magnificent, but has little real individuality of its own apart from that of Persian painting. Though transplanted to India, it had not yet begun to draw sustenance from the soil in which it had been placed. Nor should this be unexpected, rather it would appear inevitable considering the short rule, the long exile, the slow return and the immediate death of Humāyūn after he had regained the throne of Hindustan. Actually it is a matter of some surprise to notice even the small changes, the hesitating attempts towards a bolder, more expressive style which we begin to see in paintings like Khwāja ʿAbd al-Ṣamad's *Akbar presenting a miniature to Humāyūn* and *Princes of the House of Timūr,* both of which appear to have been painted in the reign of Humāyūn. And again there is no real reason to be surprised, for by his very choice of ʿAbd al-Ṣamad and Sayyid ʿAlī, who were among the most expressive and realistic artists in the Safawi context, Humāyūn was giving an indication of his own personal tastes, and also, to that extent, of the direction which the future Mughal school would take, developing these very qualities in the most amazing and vivid manner in the reign of his son the emperor Akbar.

This movement towards the distinct Mughal idiom is faintly evident, as we have shown above, in the work of Khwāja ʿAbd al-Ṣamad. But, in spite of what Abū al-Faẓl may have implied regarding his amenability to change, at least in the reign of Akbar,[103] the painter who actually seems to have made the greatest advances in the direction of the style of the new school was the incomparable Mīr Sayyid ʿAlī, his *Portrait of a young Indian scholar* (Pl. 64), whatever its date, being much closer to the style of the *Ḥamza-nāma* than any work by ʿAbd al-Ṣamad. From this point of view, his being the first to be put in charge of the great *Ḥamza-nāma* project is natural and significant, the Khwāja being entrusted with the manuscript only after seven years of work had been put in by Mīr Sayyid ʿAlī. The part played by the Mīr in the evolution of the Mughal style must have thus been considerable, and greater perhaps than is commonly realized.

The above account of painting in the reign of Humāyūn also tends to establish

103. Appendix C, p. 183.

that the *Ṭūṭī-nāma* was not a product of his reign. The style patronized by Humāyūn was primarily the well-settled Safawi style, and there would seem to be no place in its scheme for a manuscript like the *Ṭūṭī-nāma,* with its strong Indian features, restless experimentation, and clear attempts to forge a new and distinct idiom. It was not Humāyūn who gave the impetus for the formation of the Mughal style so brilliantly seen in the *Ḥamza-nāma.* That was the creation of the genius of Akbar and his times.

CHAPTER II
THE PRE-AKBAR SOURCES OF THE ṬŪṬĪ-NĀMA

Having considered the more or less Persian work done for Humāyūn, it is time now to turn to the various kinds of painting flourishing in India during and before his period. Their consideration would further amplify the context necessary for a study of the problems involving the origin of the Mughal style in the reign of Akbar, problems with the solution of which the Cleveland *Ṭūṭī-nāma* is so intimately connected. One of these Indian traditions is, of course, the well known Western Indian style, our knowledge of its richness and variety being greatly enhanced by continuing discoveries. Along with it, noticeably from the early sixteenth century, there existed other styles of Indian painting, evidence for them being once so scant that they were deemed by some not to have existed at all.[1] Even now scholarly assessments of these materials are of a preliminary nature, and our understanding has to be constantly adjusted and revised in the light of fresh materials and new interpretations. Until a knowledge of these schools becomes more secure, I propose to refer to them under the general rubric of 'pre-Akbar painting,' meaning thereby all types of painting–except the Western Indian style–done in India before the advent of the Mughal style in the reign of Akbar (as distinguished from the beginning of Mughal rule with Bābar in 1526). I prefer not to use the term 'Sultanate painting' which has been recently gaining some currency, because it implies an association with Sultanate courts with which much of the work, at least in the present state of our knowledge, does not appear to have had any direct relationship.[2]

The Western Indian style

The history of the Western Indian style is by now fairly well established. Beginning somewhere around the eleventh-twelfth centuries, it existed right up into the sixteenth century, the centre of greatest activity apparently being western India, notably the modern state of Gujarat. At least this is the impression one obtains from the large number of miniatures that have survived from this particular region. This may, however, be partly accidental, the Śvetāmbara Jaina community, particularly strong in this region, being well known for the considerable efforts it made in preserving manuscripts.[3]

Notes 1, 2, 3: see following page.

The style itself was fairly conservative, although developments are clearly discernible, and we have a wide range of achievement from the formal elegance of the early painted wooden bookcovers to the cursive and slipshod work of the sixteenth century, drowned in an indiscriminate use of blue, red and gold. A high point of the style is the splendid manuscript of the *Kalpasūtra and Kālakācarya-kathā* in the Prince of Wales Museum, Bombay, datable to the late fourteenth century.[4] In the fifteenth century, the draughtsmanship slackens though the style becomes superficially rich and elaborate, no manuscript surpassing the famous *Kalpasūtra and Kālakācārya-kathā* of the Devaśā-no-pāḍā Bhaṇḍāra at Ahmedabad, with its numerous miniatures and most elaborately ornamented borders and margins.[5] In some features, the manuscript shows an acquaintance with styles other than its own, notably work of Persian derivation being done in India, or, much less probably, Persian work proper. But what is noteworthy, is that in spite of the incorporation of several motifs and themes, the style of the manuscript itself shows no inclination to change, and in matters of line, space, form and colour, it retains the same characteristics as are to be found in any other comparable manuscript of the Western Indian style. Least of all is the Devaśāno-pāḍā manuscript affected by the Timurid style,[6] even the Sāhī figures, closest to foreign prototypes, not possessing the slightest trace of that refined colour and delicacy of line and

1. Theories of scholars who postulated the existence of pre-Akbar painting - different from the Western Indian style - without which it was not possible to reasonably understand the origins of the Mughal and Rajasthani schools, were attacked by Khandalavala, "Leaves from Rajasthan," pp. 9, 15, as an "outrage to commonsense, . . . intolerable and utterly without foundation." It is Khandalavala, however, who has been proved incorrect, the existence of these schools being now proved beyond doubt with the discovery of much fresh material. Khandalavala's attempts to date pre-Akbar painting to a post-Akbar date, often on no other ground than the occurrence in it of what he considers an exclusively Mughal item of dress, can be accounted for by his understandable desire to protect an earlier intemperate position, but are methodologically unsound and unacceptable. For literary evidence on the existence of pre-Mughal painting see Digby, "Literary Evidence," pp. 47–58.
2. Much of the material pertaining both to new developments in the Western Indian style and pre-Akbar painting has been unearthed during the last twenty odd years, and continues to grow. Most of it was initially published in the form of articles contributed to learned journals (see Bibliography). Khandalavala and Moti Chandra, *New Documents,* is a useful work bringing together much scattered material. It is nevertheless quite a confusing book for the joint authors seem to hold varying views on a large number of matters, even on small points such as the name of the Western Indian style which is also uniformly called the Gujarati style. There is not much stylistic discussion, or rather the concept of style is somewhat narrow, being confined to lengthy analyses of decorative motifs, patterns, and particularly costumes. The classifications established are also not very clear, manuscripts of the same subject being grouped together even though the style is different, and there is much loose talk of court, bourgeois, and aristocratic art which has little to do with reality but is dependent upon the author's perception of the quality of style.
3. The most comprehensive account of the Western Indian style is to be found in Moti Chandra, *Jain Miniature Paintings from Western India.*
4. Moti Chandra, "An Illustrated MS.," pp. 40–48.
5. *New Documents,* pp. 29 ff. The manuscript is datable to ca. 1475.
6. *Ibid.,* pp. 39, 41.

execution characteristic of Timurid painting. The manuscript reveals at most a willingness to look outside its own conservative milieu, but we search in vain amidst all the opulence for what can be called true stylistic change. For this, we must look elsewhere.

The Northern version of the Western Indian style

The Western Indian style, as is well known, was not just confined to Western India, but had a much larger vogue, being found in many other parts of the country, continuing researches indicating the existence of distinct regional schools. One of these seems to have flourished in northern India, its earliest known example being a *Kalpasūtra* manuscript illustrated at Yoginīpura/Delhi in V. S. 1423/1366.[7] The manuscript unfortunately has only two miniatures, but these differ from their Gujarat counterparts. In a manuscript of the *Mahāpurāṇa* of Puṣpadanta in the Śrī Digambara Nayā Mandira, Delhi, the individuality of the style stands out much more clearly. The figural types are distinct, the angularities are relatively subdued, and, there is present a new liveliness and animation. Moti Chandra assigned this manuscript to the end of the fifteenth century or a little later, and suggested a provenance in Uttar Pradesh.[8] The publication of another *Mahāpurāṇa* manuscript very similar in style, dated 1404, and painted in Yoginīpura/Delhi, suggests that the Nayā Mandira *Mahāpurāṇa* was also painted in that city, its date being nearer to the mid-fifteenth century. The discovery of several other Jaina manuscripts in a similar style, illustrating Digambara rather than Śvetāmbara texts, securely demonstrates the existence of this northern school of the Western Indian style.[9]

To this northern idiom associated with Delhi and adjacent regions can also be related a few illustrated manuscripts commissioned by Indian Muslims and written in the Arabic script, though the text itself may either be in Persian or in an Indian language like Avadhī, the fragmentary Bharat Kala Bhavan *Candāyana* being a good example.[10] There are a few changes, but these have nothing to do with style. Thus, in keeping with Islamic tradition, the format of the book is vertical rather than horizontal, but the style, as noticed by Moti Chandra, is hardly different from the Nayā Mandira *Mahāpurāṇa.*[11] The smoother drawing and greater vivacity leads one

7. Gorakshakar, pp. 56–57.
8. Moti Chandra, "An Illustrated MS. of Mahapurana," p. 79.
9. See Doshi, "An Illustrated Adipurana," pp. 382–91. A certain element of doubt regarding the date is introduced by the fact that the manuscript has only one illustration though more than three hundred were originally planned, raising the possibility of the miniature having been added at a later period.
10. Krishnadasa, "An Illustrated Avadhi MS.," pp. 66–71.
11. Moti Chandra, "An Illustrated MS. of Mahapurana," p. 80.

to suspect a somewhat later date in the second half of the fifteenth century, the provenance in all likelihood being Delhi. The *Sikandar-nāma* published by Khandalavala and Moti Chandra, almost identical in style to the Kala Bhavan *Candāyana,* is surely of the same place and period.[12] When faces are in profile, the projecting eye is shown; when they are in two-thirds profile, there is, as in the Sāhī figures no scope for the projecting eye. The dress of the male figures is also for the most part Islamic; but the figural types, the wiry draughtsmanship, the flat compositions and the colour of the two manuscripts are almost identical.

The Kala Bhavan *Candāyana* is a fragment with only a few folios, but a much better preserved and more exciting version with 140 illustrations, very similar to it in style and of about the same or a little earlier date, is in the Berlin Staatsbibliothek (Pls. 69–70).[13] The compositions of the two manuscripts are similar, but there is a verve and tension in the Berlin copy that indicates, even more than the Kala Bhavan *Candāyana* and the Naya Mandira *Mahāpurāṇa,* the vitality of this northern school. It is here that one notices an inner life that heralds change; and it is in works of this type that we will have to look for the seeds of fresh growth, not in the decoratively magnificent but essentially static and moribund style of works like the Devaśā-no-pāḍā manuscript.

A *Ḥamza-nāma* in the Berlin Staatsbibliothek[14] is also related to the Kala Bhavan and Berlin *Candāyanas* and the *Sikandar-nāma,* but seems to anticipate more clearly the style of the *Caurapañcāśikā* group in the colour, the more elaborately decorative treatment of landscape, and, to a certain extent, architecture (Pls. 71–72). As in the *Sikandar-nāma,* the background is a flat patch of red, the two-thirds view of the face is preferred, but when we do get faces in profile, the farther eye is sometimes retained, sometimes omitted, though that in and of itself is not as significant as the rather sharply reduced angularity of draughtsmanship. Fraad and Ettinghausen suggest a date of the mid-fifteenth century,[15] while Khandalavala and Moti Chandra consider it to be of the late fifteenth century and assign it a Jaunpur provenance.[16] I myself would prefer a Delhi provenance and a date around 1450–75 in view of the greater proximity in style to the *Caurapañcāśikā* group.

Our knowledge of the northern school is still very uncertain, but I would suggest as a working hypothesis the following chronology: *Mahāpurāṇa* dated 1404; Śrī

12. *New Documents,* pp. 47–50.
13. My attention to this manuscript was drawn by Dr. P. L. Gupta, its discoverer, who plans to publish it shortly.
14. Stchoukine, *Islamische Handschriften,* pp. 144–63, Taf. 9, 41–43. The manuscript has been referred to as the Tübingen *Ḥamza-nāma* from its being temporarily deposited at Tübingen, West Germany.
15. "Sultanate Painting," caption to Fig. 158.
16. *New Documents,* p. 53

Digambara Nayā Mandira *Mahāpurāṇa* ca. 1400–25; Berlin *Candāyana* ca. 1440–50; Kala Bhavan *Candāyana* and *Sikandar-nāma* ca. 1450; Berlin *Ḥamza-nāma* ca. 1450–75. The individual dates may not be correct, but the sequence of the manuscripts seems to us to be logical and worthy of credence.

The Western Indian style in eastern India

A version of the northern school of the Western Indian style also appears to have existed in eastern Uttar Pradesh and neighbouring areas. The book covers of a *Kālacakra-tantra* manuscript dated V. S. 1503/1446, painted at Arrah in Bihar, are the earliest among the few examples to be discovered, and, as is to be expected, show, together with Western Indian features, survivals of the Eastern Indian school or the Pāla style which flourished in this area before it was all but wiped out with the Muslim conquest towards the close of the twelfth century.[17] A *Karaṇḍavyūha* manuscript dated V. S. 1512/1455 is a rougher version of the same style, and is from the same general region.[18] It is the sumptuous *Kalpasūtra* painted at Jaunpur in 1465, however, which is the most important illustrated manuscript from this area.[19] The style clearly differs from that of Gujarat, being much closer to work produced at Delhi, and like it, anticipates features of the *Caurapañcāśikā* group. Previously the differences between the Jaunpur *Kalpasūtra* and other works in the Western Indian style were difficult to explain, but we can now see that one of the reasons for this difference must have been the substratum of eastern Indian traditions that existed in this region. Perhaps it was some such survival of an ancient style, of which we know presently little, that would also explain the differences between the northern version in the region of Delhi and the style as it existed in western India proper.

The Western Indian style in Malwa

Yet another important centre of the Western Indian style was Mandu, the capital of the Sultans of Malwa, a region which linked the Deccan and Gujarat to Delhi and the north. Conventional work of the type found in Gujarat was done here, but along with it, we get three very important Jaina manuscripts of exceptional quality, all datable to a little before the mid-fifteenth century, two of which have been published and are of particular importance to the study of the *Ṭūṭī-nāma.* One of these is the famous *Kalpasūtra* of 1439, now in the National Museum, New Delhi, and the other is a *Kālakācārya-kathā,* of the same date if not a few years earlier, now

17. Pal, pp. 103–111.
18. Moti Chandra, "Pair of Painted Wooden Covers," pp. 240–42.
19. Khandalavala and Moti Chandra, "An Illustrated Kalpasutra," pp. 45–54.

in the Lalbhai Dalpatbhai Institute of Indology, Ahmedabad (Pls. 67–68).[20] We again find here a clear departure from the general norms of the Western Indian style, even though some of its stock conventions like the projecting eye continue to be retained. In contrast to the rich confusion and somewhat haphazard execution, we find a studied clarity, a steadily flowing line, and a freedom and élan that leave no doubt that the Western Indian style was beginning to change in a basic and real way. We do not know the subsequent evolution of this type of work, though the presence of some of its features in the *Niʿmat-nāma* (Pls. 95–96), a manuscript of the Indo-Persian group securely datable to a period over sixty years later (1500–1510), indicates a continuity of this new idiom. The Mandu manuscripts also clearly anticipate the paintings of the *Caurapañcāśikā* group in many ways, among which can be counted the almost identical palette of bright colour, attempts at modelling the face, particularly in the regions about the nose and the mouth, the clouds, either bordered by a wavy white line, or rendered as scalloped forms filled with colour (a feature shared with the Berlin *Candāyana,* Pl. 69), and the ellipsoid tree with lanceolate green leaves veined with yellow and disposed in superimposed rows (Pl. 67). Indeed, some of these features carry over not only into the *Caurapañcāśikā* group, but also into the *Ṭūṭī-nāma,* as do several decorative motifs including the popular mauve-and-maroon chequer pattern, vermiculate designs on the borders of garments, and a characteristic ornament in which the field is divided into squares filled with floral designs, circular dots, or petalled flowers. Chevrons and petalled patterns that are to be found in the *Caurapañcāśikā* and the *Ṭūṭī-nāma* also occur here, and it is evident that we are close to the beginnings of a stylistic tradition of great consequence in the study of our manuscript.[21]

The Mandu manuscripts, even more clearly than the Berlin *Candāyana,* anticipate the style of the *Caurapañcāśīka* group. As we will shortly see, however, there is strong evidence to suggest that works of the latter group were produced not at Mandu, but in the Delhi-Agra region, from which it would seem possible to infer that the style represented by the Mandu manuscripts may have also existed in the Delhi-Agra region, or that the style of the manuscripts done at Mandu was a variation or extension of the northern school of the Western Indian style. It is only fresh evidence that can answer these questions, but the possibility that the Mandu style, as well as the styles it inspired, together with the idioms of the Delhi and Jaunpur region, had a wider geographical distribution than is at present apparent cannot be discountenanced. Our interest in these versions of the Western Indian

20. Khandalavala and Moti Chandra, "Illustrated MS. from Mandu," pp. 8–29; P. Chandra, "Unique Kalaka MS.," pp. 1–10.
21. *Ibid.,* p. 5.

style lies in the clear connection they have with paintings of the *Caurapañcāśikā* group which played such an important role in the *Ṭūṭī-nāma.*

Varieties of pre-Akbar painting

Having noticed the new developments that were taking place within the Western Indian style during the fifteenth century, we can now move on to a review of what has been defined earlier as pre-Akbar painting. The materials presently available fall into three broad categories, one of the most important being the *Caurapañcāśikā* group (Pls. 73–84), so named after the first series of paintings in that style to become known to scholars, and also posessing now two manuscripts with clear colophons stating the date and provenance, as well as several other works. The nomenclature is hardly satisfactory, but serviceable enough, and to be preferred to a designation like the "*kulaḥdār* group," based as it is on a particular type of turban which is also seen in another group of pre-Akbar paintings. The *Caurapañcāśikā* group is intimately related to the Western Indian style, especially the northern school and the Mandu version, and may be contrasted with the second or Indo-Persian group (Pls. 95–105) which is of a strikingly Persian aspect, even though Indian features are present to some degree. The corpus of materials belonging to this Indo-Persian group is growing constantly as manuscripts previously thought to belong to provincial Persian idioms are identified as being Indian. The third, or *Candāyana* group (Pls. 106–112), so named after its type manuscript, the *Candāyana* in the Prince of Wales Museum, Bombay, is related in specific features of style to both the *Caurapañcāśikā* and the Indo-Persian groups, but it is the Indian features that are to the fore, so that there is no difficulty in identifying them as Indian products, which is often not the case when we examine an Indo-Persian work. Only one more illustrated manuscript besides the Bombay *Candāyana* is known to belong to this group, and it is yet another copy of the *Candāyana* now in the John Rylands Library, Manchester (Pl. 112). The precise dating of much pre-Mughal painting remains to be accurately determined, but the weight of the evidence clearly suggests a period before the reign of Akbar, though some examples were produced after that date. A north Indian provenance for the *Caurapañcāśikā,* and perhaps for the *Candāyana* group, seems feasible; but the provenance of Indo-Persian materials is often far more indefinite.

The Caurapañcāśikā group

A study of paintings in the *Caurapañcāśikā* group has been considerably helped by the discovery of two dated examples which also state the provenance, namely a manuscript of the "Āraṇyaka Parva" of the *Mahābhārata* dated 1516 and painted at the river fort of Kacchauvā, about fifty miles from Agra (Pls. 73–74); and a *Mahāpurāṇa* dated 1540, painted at Palam, a few miles from Delhi (Pl. 75). To these may be added an etched metal bowl dated 1571, which is, however, of limited use in our study, being a provincial work in a different medium. It has been described as "a late and isolated development from styles of the kullahdar group,"[22] but does seem to give some indication of the closing phases of the style in some provincial centre.

The Kacchauvā *Mahābhārata* and the Palam *Mahāpurāṇa* suggest that the centre of the style was the Delhi-Agra region where it flourished, broadly speaking, in the first half of the sixteenth century. It would thus seem natural for it to be related to the northern school of the Western Indian style, similarities between the Nayā Mandira *Mahāpurāna* and the Palam *Mahāpurāṇa* of 1540 having been noticed already by Moti Chandra.[23] To these may be added the even closer stylistic affinities between paintings of the *Caurapañcāśikā* group and the Berlin *Ḥamza-nāma* (Pls.71–72), another work related to the northern version of the Western Indian style, in matters of colour, conventions for treating landscape, and ornament. It would thus not be improper to suggest, on the basis of evidence presently available, that the *Caurapañcāśikā* group not only followed, but in some sense evolved from the northern school of the Western Indian style some time around the opening years of the sixteenth century. It could conceivably have come into existence even earlier because of its close affinities to the Mandu *Kalpasūtra* of 1439 and the related *Kālakācārya-kathā* of about the same date, this relationship raising the possibility of the style of the group extending into Malwa. Evidence for its extension further east into the Gangetic plain is unclear, but relationships with the Jaunpur *Kalpasūtra* of 1465 indicate that this may well have been the case.

The number of paintings and manuscripts in this style were once few, but the corpus is now fairly large. Several paintings, almost identical in style to the Kacchauvā *Mahābhārata* of 1516, are to be found in a profusely illustrated *Mahāpurāṇa* of the Sagar Bhaṇḍāra.[24] The Palam *Mahāpurāṇa* of 1540 (Pl. 75) is a rather simple production, showing little stylistic development; and, had it not been for the clear date, it could well have been placed at no great distance from the Kacchauvā *Mahābhārata.* The *Miragāvata* of the Bharat Kala Bhavan,[25] and a

22. Digby, "Bhugola of Ksema Karna," p. 19. The place where the object was made is not given.
23. Moti Chandra, "Ilustrated MS. of Mahapurana," p. 80
24. I am thankful to Dr. Sarayu Doshi for bringing the manuscript to my notice. It has in the main paintings by two groups of artists, one of them related to the northern Indian school of the Western Indian style, and the other to the *Caurapañcāśikā* group, executed in a style close to the 1516 *Mahābhārata* but of greater vitality. The juxtaposition of these two groups of work may perhaps be of significance.
25. Ananda, Krishna, "Some Pre-Akbari Examples," pp. 18–22; and Khandalavala, "Mṛigāvat of Bharat Kala Bhavan," pp. 19–36.

Rāgamālā series formerly in the collection of Vijayendra Sūri[26] also present many similarities to the Kacchauvā *Mahābhārata* and the Palam *Mahāpurāṇa,* and can be dated to the same period, ca. 1520 to 1540. The *kaithī* script of the *Miragāvata* suggests a provenance nearer Jaunpur and eastern Uttar Pradesh, and though this seems possible, the evidence is not conclusive.

Problems in dating the paintings cited above are relatively few considering the fairly close similarities and the relatively simple and modest execution.[27] But the situation is more complex when we consider the principal glories of the *Caurapañcāśikā* group: the thoroughly dispersed manuscript of the *Bhāgavata-purāṇa* (Pls. 76–79), the *Caurapañcāśikā* series itself (Pls. 81–84), a manuscript of the *Candāyana* now divided between the museums at Lahore and Chandigarh, and a fragmentary *Gīta-Govinda* series in the Prince of Wales Museum of Western India (Pl. 80).[28] Most likely these paintings also belong to the same period as the *Mahābhārata* and *Mahāpurāṇa* (ca. 1516–1540), their more assured and accomplished character to be explained by the workmanship of superior artists in the service of more demanding patrons. It has also been suggested that these paintings are even earlier, the Kacchauvā *Mahābhārata* and related work being popular and simplified derivations from them, a point of view worthy of consideration in view of the close affinities to the Mandu *Kalpasūtra* of 1439. The likelihood of their being later, the result of Mughal influence on a manuscript like the Kacchauvā *Mahābhārata,* does not seem possible for the simple reason that there is no Mughal influence observable in these examples. The mere presence of the *cākadāra-jāmā,* which figures so largely in the arguments of Khandalavala is of no significance,[29] for a type of dress cannot be symptomatic of a style.

Paintings of a post-Akbar phase but in the style of the *Caurapañcāśikā* group are now known, and the presence of certain Mughal features in them, and the absence of these very features in the paintings referred to above, further strengthen a belief in

26. Norman Brown, pp. 1–10.

27. *New Documents,* p. 69, refers to these works as a "school of bourgeois painting." Aside from the fact that the patrons of the 1516 *Mahābhārata* seem to have been officials belonging to a petty, landed aristocracy *(caudharī)* and not merchants, it is improper to assign patronage of works of art to a particular section of society solely on considerations of quality. There is little to indicate that the "bourgeoisie" did not patronize the other more elaborate works of the *Caurapañcāśikā* group, and we know that merchants caused to be made some of the most magnificent works executed in the Western Indian style.

28. Scattered pages of the *Bhāgavata-purāṇa* are found in numerous museums and private collections. A few have been published in *Heeramaneck Catalogue,* p. 101, *Rajput Miniatures from the Collection of Edwin Binney,* pp. 4–5, and *New Documents,* p. 84, Plates 199–200; for other paintings of this group see Shiveshwarkar, Tarafdar, and *New Documents,* Plates 188–95 and Colour Plates 22, 23.

29. His most recent statement on the subject is in "The Mṛigāvata," pp. 30 ff. He fails to see that it is not the presence or absence of the *cākadāra jama* but the manner in which it is painted that should be of concern and significance to the historian of art.

the pre-Akbar date of the main *Caurapañcāśikā* group (see *infra* and p. 41). The appearance in the Cleveland *Ṭūṭī-nāma,* in our opinion a very early Akbar period manuscript, of several features derived from the *Caurapancāśikā* group would also seem to preclude a post-Akbar dating. All things considered, we find a date in the first half of the sixteenth century acceptable, though the possibility of an earlier date has to be kept open.

Of the examples mentioned, the *Bhāgavata-purāṇa* (Pls. 76–79) is most like the Kacchauvā *Mahābhārata* of 1516 in its free line and facile rendering. The work is spontaneous rather than studied, the compositions often of great originality and complexity, the movement lively and vivacious so that we get the impression of a style still in the making. For these reasons, I think of the *Bhāgavata-purāṇa* as being stylistically anterior to the *Caurapañcāśikā* and the Lahore and Chandigarh *Candāyana,* which are magnificent works of greater elegance, but controlled, deliberate and precise in their execution, characteristics one would expect of the style in its mature and well-settled phase. The Prince of Wales Museum *Gīta-Govinda* is not unlike the *Caurapañcāśikā,* but is less rigorous in execution and is also imbued with considerable emotional content as in the depiction of the pining Rādhā (Pl. 80). The faces are also different, more rounded, with receding foreheads; and the blooming landscape is of unusual beauty.

It is only with the Issarda *Bhāgavata-purāṇa,* now dispersed (Pls. 85–86),[30] and a few folios from a *Gīta-Govinda* series in the National Museum of India[31] that we first begin to see Mughal influence on the *Caurapañcāśikā* group, most notably in the finer line, a more minute and painstaking rendering of detail, a greater depth in the composition, more subdued colour, and marked delicacy of execution. In matters of detail also, notably the swirling treatment of water, (cf. Pls. 78, 79 and Plate 85) and a curious tree, the foliage of which reminds one of superimposed lumps of rock sprouting leaves at the base (Pls. 85, 86), we get glimpses of workmanship that has been modified by Mughal experience. These influences, nevertheless, hardly alter the basic nature of the style, which retains its integrity and is able to assimilate Mughal features successfully. Even stronger Mughal influences on the *Caurapañcāśikā* group, absorbed equally if not even more successfully, are to be seen in a *Bhāgavata-purāṇa* discovered in Ahmedabad and to be published shortly (Pls. 87–90).[32] The compositions are similar to those of the Issarda *Bhāgavata,* but Mughal influence is

30. The manuscript is so named because it was formerly in the possession of the chief of Issarda, a vassal of the Maharaja of Jaipur. For reproduction see Spink, Fig. 11, and Welch, *Meadow,* p. 26.

31. *New Documents,* Pls. 201–202.

32. I am indebted to Dr. Ratan Parimoo for bringing these paintings to my attention and it is through his kindness that the photographs are reproduced in this book. Dr. Parimoo's article will be published in *Lalit Kala,* No. 17.

even more apparent in the livelier movement, the smoother draughtsmanship, a keener psychological perception, and a more evident rendering of depth. The swirling eddies of water are obviously derived from the *Ḥamza-nāma* idiom (Pl. 87), as are the trees with shaded trunks (Pl. 88), the swaying springy fronds of palms (Pl. 89), and the shaded rocks (Pl. 88 .) Architecture also is shown in some depth, and the slender columns with herring-bone pattern are Mughal in origin (Pls. 89–90). The human figures are not as stiff, and are often fairly relaxed and more fully modelled; and the conventions for depicting folds of cloth, or giving it a gauzy texture (Pls. 88, 89), also show a Mughal source. The agonized expression of Kāla-yavana, his body bursting into flames, the teeth bared in a painful grimace, would not be possible without Mughal intervention (Pl. 88). At the same time, what has been borrowed has been skilfully integrated. The rich geometrical and floral ornamentation of architecture seen in the Mughal style has not been attempted; and even in the rendering of trees and the shaded rocks, an appropriate simplification has been introduced.

We have considered in some detail the style of these late *Caurapañcāśikā* group manuscripts (though they are of a period after the establishment of the Mughal style and could thus not have contributed to its formative phases), in order to demonstrate the nature of changes that occurred in the pre-Akbar styles when they did come in contact with the Mughal idiom. It has been suggested by some scholars that the *Ṭūṭī-nāma* is an example of a pre-Akbar style under Mughal influence and, if this were indeed so, it should have possessed at least the general character of the Ahmedabad *Bhāgavata;* but this is emphatically not the case, as can be seen by comparing it with fol. 43r of the *Ṭūṭī-nāma,* a miniature with strong *Caurapañcāśikā* affiliations. That miniature represents rather an opposite process in which a work deriving from the *Caurapañcāśikā* group is undergoing a process of Mughalization, and is quite different from manuscripts like the Ahmedabad *Bhāgavata-purāṇa* which show the style of the *Caurapañcāśikā* group at the receiving end. Another series showing the style of the *Caurapañcāśikā* group under Mughal influence is a *Rāgamālā* in the Bharat Kala Bhavan (Pl. 91), not far removed in style from the Issarda *Bhāgavata-purāṇa.*[33] Even stronger Mughal features are to be seen in some folios of a *Bhāgavata-puraṇa* in the Kankroli collection (Pls. 92–93). Here, the nature of the style itself has undergone considerable transformation, only a few vestiges of the *Caurapañcāśikā* group, notably the red background, the mauve-and-maroon chequer pattern, and the wavy skyline at the top, surviving.

The precise chronology of the Mughal-influenced phase of the *Caurapāñcaśikā* style remains uncertain, but it is reasonable to assume that the manuscripts described

33. Mukherjee, Pl. XXVIII.

were painted subsequent to 1570 after the *Ḥamza-nāma* style had become well established. The Issarda and the Ahmedabad *Bhāgavata-purāṇas* referred to above can be thus tentatively dated to ca. 1575–80, the Kala Bhavan *Rāgamālā* shortly thereafter, while the Kankroli *Bhāgavata* is likely to have been painted in the last decade of the sixteenth century.

To recapitulate, the style of the *Caurapañcāśikā* group seems to have come into existence some time around the opening years of the sixteenth century. The present evidence seems to indicate that it flourished in the Delhi-Agra region, though a wider geographical provenance is possible. The Kacchauvā *Mahābhārata* of 1516, the *Miragāvata,* the Palam *Mahāpurāṇa* of 1540, and the *Caurapañcāśikā,* the dispersed *Bhāgavata-purāṇa,* the Punjab Museum *Candāyana,* and a few other works were painted in the first half of the sixteenth century, before the advent of the Mughal style. As such, they are particularly important for a study of the *Ṭūṭī-nāma,* and constitute one of its sources. The style of the group began to be influenced by the Mughal school about 1575; and by the close of the century, it had succeeded in substantially altering its character. The later history of the *Caurapañcāśikā* group is unknown, though it seems to have ceased to exist as an individual entity, its traditions probably absorbed and continued by the rising Rajasthani schools of the seventeenth century.

Indo-Persian painting

Both the Western Indian style and the style of the *Caurapañcāśikā* group are purely indigenous, but along with them there existed illustrated manuscripts of a much more exotic character, constituting what is called here the Indo-Persian group. The miniatures in these manuscripts are derived from a variety of Persian styles or a combination of them, and so removed are they in outward aspect from Indian work that they were commonly considered by scholars of Persian painting as belonging to provincial Persian traditions though they were unable to place them successfully in any firm Persian context. It is only recently that convincing grounds have been advanced for assigning illustrated manuscripts of this type to India, particularly through the efforts of Goetz, Welch, Skelton, Robinson, and Ettinghausen, all of whom have the advantage of being equally learned in Indian as well as Persian painting.[34]

34. Goetz, "Vestiges of Muslim Painting," p. 212 had suggested the possibility of assigning an Indian origin to manuscripts attributed to unidentified Persian studios. Fraad and Ettinghausen, pp. 48–66; Welch, *Meadow,*" pp. 86–87, fn. l; and Binney, *Catalogue Mughal and Deccani Schools,* pp. 15–22, 136–40, provide additional information.

Even now, the line separating what is provincial Persian work from work produced in India can be very fine. Nor has it been possible to identify schools, assign provenances and trace patterns of development, so that the impression we presently get for the most part (and which may have to be modified with further discoveries) is of an *ad hoc* output, rather than a consistent style, produced by and large by Indian artists imitating or inspired by a medley of Persian work, or by Persian artists and their followers working in India, their paintings impressed to a lesser or greater extent with a somewhat changed aspect in the new environment. The patrons appear to be Muslims, probably the nobility, gentry and scholars. Only a few manuscripts are known to have been made for royalty, and it is unclear to what extent this kind of work was patronised by the Sultans of Delhi or of the provincial kingdoms, though they must have certainly done so. It is also evident that Indo-Persian manuscripts were done both in the north and in the Deccan, and began to be produced as early as the first decades of the fifteenth century if not earlier. Miniatures in this manner continued to be painted upto the sixteenth century, and vestiges can be seen in Indian styles of an even later period. It is to paintings of the Indo-Persian group, rather than to Persian painting proper, that several Persian features noticed in the Western Indian style and the *Caurapañcāśikā* group can be properly attributed. Several features in the Cleveland *Ṭūṭī-nāma* are also derived from works of the Indo-Persian group. Their share in the development of the Mughal style must have been much greater than is realized, and, for all we know, some kind of Indo-Persian work may have even been practiced at the Mughal court before Akbar. In any case, paintings of this type had been produced in India from at least the fifteenth century, and probably much earlier, along with the Western Indian and other indigenous styles; and artists practicing this manner must have contributed to the shaping of the distinctive Mughal idiom, together with the new Safawi masters recruited by Humāyūn. A painter like S͟haḥm Muẕahhib, working in a marked Indo-Persian manner of Bukhara derivation, whose paintings are to be seen in the British Museum *Gulistān* (Pls. 37–38)[35] and also the 1570 *Anwār-i Suhailī* in the School of Oriental Studies, London (Pl. 42), both of the Akbar period, could equally well be an Indian painter inherited by the Mughal atelier either from a provincial centre where Indo-Persian work was patronised, or for that matter, from Humāyūn's atelier itself, rather than an artist directly recruited from Bukhara. Be that as it may, what is increasingly evident with every new discovery is the considerable corpus of Indo-Persian work that existed for the Mughal school to draw upon, both with regard to style and artists; and these had as influential a part to play

35. Reproduced in Martin, Plates 146–47.

in the development of the Mughal school as any of the other pre-Akbar styles we have drawn attention to earlier.

Some idea of the variety of Indo-Persian painting can be had from the examples studied by Fraad and Ettinghausen, primarily of the first half of the fifteenth century,[36] the *Būstān* of Saʿdī in the National Museum of India,[37] the India Office Library *Sindbād-nāma* and the Victoria and Albert Museum *Anwār-i Suhailī*, both of ca. 1550,[38] and the *Arbaʿah* of Hātifī which may be a little later.[39] These provide us with impressive documentation of the type and tradition of work that could have contributed to the Mughal school. Here, I will content myself by describing a manuscript of the *Khamsa* of Niẓāmī in the collection of Seth Kasturbhai Lalbhai, a rather good example of the better class of manuscripts, and of particular interest to us in as much as it clearly demonstrates the relationship between the Indo-Persian and the Mughal schools.[40]

Seventeen miniatures of a total of thirty-four in the manuscript are in a pronounced Persian style with hardly any noticeable Indian features. Of these, eleven, including seven which are repetitious representations of Bahram Gur in pavilions of various colours, are in a style closely resembling that of Bukhara between ca. 1525 and 1550, except for an occasional touch of bright orange red or vivid yellow that betrays an Indian palette. Plate 100 is the best example of this type. The other six seem to be done primarily in a manner recalling the Safawi style of Tabriz as practised at the court of Shāh Ṭahmāsp between 1525 and 1540. Plates 97, 98 are good examples of this kind of work, though they possess none of the rich and sumptuous execution of comparable Persian masterpieces, being somewhat impoverished representations thereof. The presence of these Persian-type miniatures both of Tabriz and Bukhara derivation in the same manuscript is in itself interesting, this mixing up of styles derived from different Persian sources being characteristic of several Indo-Persian manuscripts. The date of the manuscript, based on the Persian type miniatures, would appear to be somewhere between 1525 and 1550,[41] probably before the reign of Akbar, but still close enough to it so as to

36. "Sultanate Paintings," pp. 48–66.

37. Ettinghausen, "The Bustan MS." pp. 42–43.

38. Stchoukine, *Manuscrits Safavis*, pp. 137, 138, Plates 78–79 and 85–86.

39. Arberry, *Beatty Persian Manuscripts Catalogue* III, pp. 35–36.

40. The manuscript is approximately 27.5 x 18 cms., the area of the text covering 21 x 10.8 cms. It contains thirty-four miniatures, each one of them measuring approximately 15 x 10.8 cms. respectively. There is no colophon or indication of date. The double title-page is exquisitely illuminated, and carries verses in praise of Niẓāmī. The writing is a very fine *nastaʿlīq*.

41. The Bukhara type miniatures are closely comparable to those found in a manuscript of the *Būstān* of Saʿdī dated ca. 1525–30 (Royal Asiatic Society, Morley 251, reproduced in Robinson, *Persian Painting*, Fig. 18) and a *Sabʿa Sayyāra* of Nawāʾī dated 1553 (Bodleian Library; Robinson, *Descriptive Catalogue*, No. 978, Pl. XXII). Miniatures resembling the Safawi style of Tabriz are derived from types found in the *Shāh-nāma* of Shāh Ṭahmāsp datable to ca. 1525–40, see Welch, *King's Book of Kings*.

make the possibility of contact with the Mughal school evolving under his patronage quite a real one.

The third type of illustration in the manuscript consists of fifteen miniatures of Indian workmanship (Pls. 99, 101–105). Some of these closely parallel the more Persian work, but can be nevertheless relatively easily recognized; while yet others have a fairly pronounced Indian aspect revealed in the marked tendency to compose miniatures in terms of fairly distinct registers; landscape elements such as plantain trees, palms, and arabesque clouds; architectural elements such as domes, pillars, and embattled parapets; the treatment of animals, notably the black antelope; or details such as crowns, flywhisks, and ceiling ornaments. Several faces shown in profile are distinctly Indian (Pl. 99); and the garments are often decorated with spotted and squiggly decoration (Pl. 101). The colours are much warmer in tone, and the line harder and more hesitating. Some of these features are shared with the *Candāyana* group, which is of a much more obviously Indian aspect, and reinforces our belief that these also are works of Indian artists.

Plates 97 through 105 represent a cross-section of the manuscript and give a good idea of the nature and range of the work. *Nau͟shīrwān and the owls* (Pl. 97) is among the most Persian, and is very similar to the style of Safawi work seen, for example, in a representation of the same scene in the British Museum *͟Khamsa* of Niz̤āmī painted between 1539 and 1543 (Binyon, *Poems of Nizami,* Pl. III), though the Indian version is hardly as magnificent, the landscape in particular being much less colourful. A battle scene (Pl. 98) is also very Persian, with hardly a trace of Indian features. By contrast, Plate 99, representing a combat between two horsemen, though derived from the Safawi style as seen in some miniatures of ͟Shāh Ṭahmāsp's *͟Shāh-nāma* (Welch, *King's Book of Kings,* p. 165), possesses many strikingly Indian features, notably the rather strict horizontal alignment of figures, the flat strips of curving blue and gold sky, and, most strongly, the drawing of the faces of the group of soldiers behind the hillock to the left, as well as that of the unhorsed cavalier. These are all shown in profile, and are only mildly Persianised versions of the kind of faces seen in the 1516 *Mahābhārata* or the 1540 *Mahāpurāṇa* (cf. Pl. 74). The court scene illustrated in Plate 100 is, as we have stated earlier, strongly dependent upon Bukhara traditions; while Plate 101, though following this type of work, is much more strikingly Indian in drawing as well as in colour. A miniature in the Cleveland *Ṭūṭī-nāma* is closely connected to work of this type (fol. 339r). Plate 102 depicting the meeting of Lailā and Majnūn is an attractive miniature, the ground covered with flowering plants. Beyond the red fence in the background are plantain trees and swaying palms of emphatically Indian character, as is the flywhisk and the pointed crown worn by the lady. Plate 103 showing a man playing music in the company of animals has a curving horizon and sky in blue and gold strips similar to Pl. 99. Once again the black buck and the composition in emphasised registers are quite Indian, and the basis of several *Ṭūṭī-nāma* miniatures (fols. 16r, 26v, 152r). Plate 104 is probably by the same artist as Plate 101, and is painted in the same tones of colour. The division of the plane into two registers by a narrow band decorated with arabesques, as well as the compartmentalisation of the lower register into arched panels are Indian features seen in the Bombay *Candāyana,* where we get the gold arabesque clouds on blue as well (Pl. 107). The painting with the most pronouncedly Indian features in the Lalbhai *Khamsa,* however, is *Majnūn lamenting at the tomb of Lailā* (Pl. 105). The composition is divided into three clear registers with a pool and shrubs in the foreground, the main scene, where the emaciated Majnūn is shown clutching at the grave, being placed in the centre. The entire arrangement with plain monochrome colour of the wall, the ornament suspended from the ceiling, the embattled parapet and the three domes is strikingly reminiscent of *Candāyana* group paintings (Pls. 107, 108). A painting of this type strongly influenced folio 128r of the Cleveland *Ṭūṭī-nāma,* the similarities being most striking.

While the date of the Lalbhai manuscript can be placed with some confidence between ca. 1525 and 1550, it is difficult to determine the provenance. Affinities to the *Candāyana* group suggest north India and not the Deccan. Malwa remains a possibility, for it seems to have been an important centre of Indo-Persian work, but there it little evidence to support a definite assignment to that place. The quality of the manuscript, though not the most outstanding by Iranian standards, is not less than, and is perhaps superior to that of the Mandu *Būstān* produced for Nāṣir Shāh Khaljī, so that fairly exalted, even royal patronage is plausible.

Provenances of manuscripts belonging to the Indo-Persian group are, as stated previously, quite uncertain, but we get a few examples known to be from Mandu, which has already been observed to have been an important centre of a progressive Western Indian school. An Indo-Persian style based upon the Turkman idiom of Shiraz is also to be seen there during the last quarter of the fifteenth century in a manuscript of the *Miftāḥ al-Fuẓalā,* a glossary compiled by Muḥammad of Shadiabad/Mandu. The date of the work is A. H. 873/1468–9,[42] and the manuscript does not appear to be much later judging from Turkman work proper of that period. Though Indian features are not very obvious, the several misunderstood Persian conventions reveal it to be the work of Indian artists.

Indian elements in the well-known *Niʿmat-nāma* (ca. 1500–10) made at Mandu are much clearer, these being derived from the kind of work seen in the Mandu *Kalpasūtra* of 1439 (Pls. 95, 96), while the Persian features are a continuation of

42. Titley, pp. 15–19. There is a discrepancy in the date given, A. H. 873 being equivalent to 1468–9 and not 1471–2.

characteristics seen in the *Miftāḥ al-Fuẓalā.*[43] The *Niᶜmat-nāma* paintings thus are much more obviously by Indian painters with Turkman training or under Turkman influence, and this type of work together with other pre-Mughal idioms must have also made its contribution to the evolution of the Mughal school when it came into being early in the reign of Akbar.[44] It has been suggested by Khandalavala that the Persian type of human figure in the *Niᶜmat-nāma* is so Persian that it is not possible to seriously countenance an Indian painter for it,[45] but this argument is quite fallacious, and one might as well argue that the Indian type of figure is so Indian that a Persian painter could not be seriously considered. Rather, one has to agree with the well-considered opinion of Robert Skelton who has drawn our attention to the more conceptual approach of the *Niᶜmat-nāma* artists and the mistranslations of Persian conventions, thus providing valid stylistic criteria for his assessment of the painters as being trained in the Western Indian style of Mandu[46] and their subjection to Persian discipline that almost totally changed their style. A similar process is once again seen operating in the *Ṭūṭī-nāma* where a variety of work, including, perhaps, the type seen in the *Niᶜmat-nāma,* was being coaxed to assume the dimensions of a newly developing Mughal style. Khandalavala's argument that a painter of the *Niᶜmat-nāma* type could have hardly been acceptable to the Mughal atelier when it possessed so great a master as Mīr Sayyid ᶜAlī is to be quickly dismissed,[47] for if this were true, the Mughal atelier could not have grown beyond the few Persian masters that came to it in the reign of Humāyūn and Akbar. No painters of the Indo-Persian group would have been recruited, even though they worked in a style that, of all Indian work, would have struck a sympathetic chord in the Safawi masters; and, as for artists working in other pre-Mughal traditions, they would never have been able to get in, so removed was their workmanship from the refined luxury of the Safawi idiom. On the contrary, the style of Mughal paintings, the names of painters as preserved for us in contemporary attributions, and literary testimony make it clear that Mughal artists were drawn from an extensive geographical area, possessed varying talents, and belonged to different stylistic traditions. Conformity to the

43. See P. Chandra, "Notes on Mandu Kalpasutra," pp. 51–54. The National Museum *Būstān* dated ca. 1502–03 though painted at Mandu is not in the style of the *Miftāḥ al-Fuẓalā* and the *Niᶜmat-nāma,* being dependent rather on a provincial Herati manner. From this it would appear that works other than those in the local style, if any, were also encouraged provided a painter was available, and tends to confirm the heterogenous nature of much Indo-Persian work.
44. Skelton, *Niᶜmat-nāma,*" p. 48, had already noticed odd details in the work of Akbar's artists anticipated in the *Niᶜmat-nāma* though at the time he wrote the connection between Indo-Persian painting and the Mughal style was still obscure. The *Ṭūṭī-nāma* clearly shows this connection.
45. *New Documents,* p. 61.
46. Skelton, "Niᶜmat-nāma," p. 48.
47. *New Documents,* p. 63.

grand Safawi manner was certainly no criteria for recruitment, nor could it be, for the Mughal style never made an attempt to be its copy, and was actually setting for itself, under the patronage of Akbar whose love of things Indian is too well-known to need any demonstration, norms and ideals that would result in the development of work of a character quite different from Safawi achievement.

The Candāyana group

The Western Indian style together with that of the *Caurapañcāśikā* group can be effectively contrasted with paintings of the Indo-Persian group even though they do share a few features in common, both being, after all, Indian styles. This dichotomy, however, already eroded in the *Niᶜmat-nāma,* is even more effectively done away with in what may be called the *Candāyana* group of paintings so named after an illustrated manuscript of that work in the Prince of Wales Museum of Western India, Bombay (Pls. 106–111).[48] Here the compositions and the human figures shown in profile and in Indian dress are derived from the *Caurapañcāśikā* group. The subdued and pale colour, if not the manner in which it is used, the finer line and the decorative motifs, including conventions for clouds and the ever present arabesque, are, however, of Indo-Persian derivation, and to some extent dependent upon manuscripts like the *Niᶜmat-nāma.* Another *Candāyana* manuscript in the Rylands Library, Manchester (Pl. 112) is similar in style to the Bombay *Candāyana,* though of rougher workmanship.[49] Its date, as well as that of the Bombay manuscript, has not yet been accurately determined, but they both would seem to belong to the first half of the sixteenth century, the Manchester manuscript probably being the later of the two. A date after the inception of the Mughal school is as unlikely as it is for paintings of the *Caurapañcāśikā* group, and for the same reasons.[50] The provenance is also uncertain, and it is entirely possible that works of this type may have had the same geographical distribution as the *Caurapañcāśikā,* group. The *Candāyana* group is of the greatest importance to us as painters practicing its style played an important part in the execution of the *Ṭūṭī-nāma.*

This brief account of pre-Akbar painting should have made evident the rather loose understanding we have of the period and the tenuous nature of even the broad and general conclusions that we have attempted to reach. Nevertheless, the existence, aside from the Western Indian style, of a fairly rich variety and amount of

48. What we call the *Candāyana* group and the manuscript after which it is named must not be confused with the *Candāyana* manuscripts in the northern version of the Western Indian style or the *Candāyana* manuscript of the *Caurapañcāśikā* group.

49. *New Documents,* Colour Plate 25, and Plates 176–77, pp. 99 ff.

50. See *supra,* p. 39. Moti Chandra, *New Documents,* p. 94 dates the manuscript to ca. 1525.

work practiced throughout north India and also in the Deccan, ranging from strongly Persianising work of the Indo-Persian group to the style of the *Caurapañcāśikā* group, which is essentially indigenous and very different both in concept and execution is clear and irrefutable. When the Mughal school came into existence in the reign of Akbar, it did not spring into being out of nothing. There was a fairly rich heritage and a widely diverse repertoire from which it could draw, and that it indeed did so is now quite clear. What we have to remember though is that what was taken from previously existing traditions was transformed in such a quick and revolutionary manner that the connections have often been felt rather than clearly seen. The Cleveland *Ṭūṭī-nāma* is the one manuscript hitherto discovered that not only clearly demonstrates this connection between the pre-Akbar traditions and the Mughal style, but also reveals the actual processes by which the varied and disparate artistic elements of pre-Akbar work were being thoroughly transmuted, ultimately evolving into a quite new and homogenous style of high accomplishment and power, quite distinct from the various idioms that contributed to its birth. The Rajasthani schools that began to come into existence towards the late sixteenth century, on the other hand, represent no such process of drastic transformation. They continue, rather, in a clear and directly perceptible manner mainly the traditions of the Western Indian style and the *Caurapañcāśikā* group, with a leavening of Mughal influence which was again absorbed without affecting in any way the essential nature of the style itself. Paintings of the Chawand *Rāgamālā* are particularly illustrative, being still fairly close to pre-Akbar idioms (Pl. 94).[51]

51. The series are supposed to date, according to an inscription on one of the miniatures to V. S. 1662/1605, though stylistically, as suspected by Mr. Welch, they seem to belong to an earlier phase. It may perhaps be possible to suggest that the date has not been accurately reconstructed for the letters are quite defaced, see Kanoria, Colour Plate facing p. 1. It is not impossible to read V. S. 1632/1575 which is more satisfying from the point of view of style. Chawand must have also gained in importance around this time, Chittor having fallen in 1567. According to Śyāmaladāsa, *Vīra-vinoda,* p. 159, Mahārāṇā Pratāpa made it his capital sometime around 1576–80.

CHAPTER III
AKBAR AND MUGHAL PAINTING

The reign of Akbar

The future emperor Akbar, one of the greatest sovereigns to rule India, was born on October 15, 1542 at a time when his father Humāyūn's fortunes were dangerously low, for he had been reduced to the life of a homeless wanderer after suffering defeat by Sher Shah Sūr. When Humāyūn finally decided to seek refuge in Persia, Akbar, barely a year old, had perforce to be left behind, falling first into the custody of his uncle Mirzā ʿAskarī who was in charge of Kandahar, and later into that of yet another uncle Mīrza Kāmrān who governed Kabul; and it was only after the capture of that city by Humāyūn in 1545 that he was reunited with his parents once more. Of Akbar's life in Kabul during the nine years his father was established there, we know little, except that he did not take to conventional studies, and that he had to undergo the horrifying experience of being exposed to his father's guns on the ramparts of Kabul when the city was temporarily captured by the rebellious Mirzā Kāmrān. It also seems that Akbar took lessons in painting, this being established by a miniature done by Khwāja ʿAbd al-Ṣamad in which two youths are shown in a landscape, the one working with the brush being the young Akbar.[1] Akbar accompanied his father when he returned to Delhi in July 1555, and was with him until November of that year, when he was nominally appointed to the Punjab under the guardianship of the redoubtable Bairām Khān. There is a very clear reference to Akbar practicing painting while he was at Delhi by Abū al-Faẓl. According to him, Akbar was being instructed in the art by skillful artists such as Mīr Sayyid ʿAlī and Khwāja ʿAbd al-Ṣamad, and one day while he was practicing in the library, he drew a picture of a man with all the limbs separated. When asked by a courtier as to who this was, Akbar replied that it represented Hemū. The young emperor remembered this incident when he declined the dubious honour of becoming a *ghāzī* by refusing to kill the captive Hemū with his own hands a year later, and was himself the authority for Abū al-Faẓl's account.[2]

Akbar ascended the throne at Kalanur in the Punjab in February 1556 after the accidental death of Humāyūn just six months after his return to India. He did not

1. BWG, Pl. CV-B; also see *supra*, p. 22.
2. *Akbar Nāmā* II, p. 67. Jahāngīr also refers to the same incident except that he mistakenly places it in Kabul, and adds the information that Akbar was working in Khwāja ʿAbd al-Ṣamad's presence, see *Tūzuk* I, p. 40.

return to Delhi immediately, the way being barred by the valiant Hemū, a Hindu general of the Afghans who had taken advantage of the death of Humāyūn to capture Agra and Delhi. Fortunately, he was defeated at the Battle of Panipat in November 1556, and Delhi and Agra were reoccupied, but it was the end of 1558 before Akbar entered Agra to settle down there. He soon fell out with Bairām Khān who was dismissed, defeated, and ordered off to Mecca, on the way to which, he was assassinated by some revengeful Afghans. These events occurred in 1560 and early 1561, and by 1562, at the age of twenty, Akbar had begun to exercise full power. This was also the year in which he seems to have gone through a deep spiritual experience[3] which may account for orders prohibiting the enslavement of Hindu prisoners of war (1562), the throwing open of state service to Hindus, the abolition of the pilgrim tax (1563), followed immediately by the repeal of the hated *jizya* (1564), the poll-tax levied on non-Muslims. The young emperor in his very first acts of state broke sharply with the past, and adopted policies that in no small way contributed to the great achievements of his reign.

During these years, Malwa was also conquered (1560–62) and Jaunpur subdued (1561), the young emperor being fully occupied with military and political activities; but the arts were hardly neglected, and the famous musician Tānasena was sent to the court by Rājā Rāmachanda of Rewa (1562). Akbar also began his great architectural activities around this time with constructions at Agra fort, culminating in what amounted to the erection of a new fort in A. H. 972/1565 which was completed in about A. H. 986/1578–79.[4] While this immense undertaking was still in progress, work was also begun on the new capital at Sikri in A. H. 976/1568 and completed in A. H. 981/1573,[5] though royal buildings must have continued to be made as witness the Ibādat-khānā put up in 1575, and the Buland-darwāzā dated A. H. 983/1575–76.[6] Several buildings at Fatehpur-Sikri, as the city has come to be called, still retain the paintings with which they were once decorated, notably the Khwābgāh and Mariam's House, but these are in a sadly mutilated condition. Though the buildings are themselves not dated, it is reasonable to assume that they were built between 1568 and 1573.[7]

Fatehpur-Sikri became Akbar's residence until 1585, and was the scene of great accomplishments. The powerful Rajput states were reduced to submission, and the brave Mahārāṇā of Mewar driven into the wilderness. Bihar and Bengal were brought

3. *Akbar Nāmā* III, p. 386.
4. Nur Bakhsh, pp. 164 ff.
5. This date is according to the chronograms given by Badāyūnī II, pp. 112–13. *Akbar Nāmā* II, p. 530 gives the impression that serious work began in 1571 but the foundations had been laid earlier.
6. *Akbar Nāmā* III, p. 157. The date of the Buland-darwāza is based on a traditional chronogram which appears to be reliable.

Note 7: see following page.

under Mughal sovereignty. Gujarat was conquered, and the wars there provided opportunities for direct contact with the Portuguese (1573). This may have inspired the dispatch of Ḥājī Ḥabībullāh of Kashan to Goa with orders to bring back the wonderful things of that country, and also to take clever craftsmen with him so that they may learn new skills. This the Ḥājī did, and Abū al-Faẓl describes his return with a large number of persons dressed as Christians and playing European drums, craftsmen displaying the skills they had acquired, and musicians playing on an organ.[8] A request was subsequently sent to Goa for the dispatch of learned Christian priests, and resulted in the arrival of the first Jesuit mission at the court in 1580 headed by Rudolfo Aquaviva. He carried with him a copy of Plantyn's *Royal Polyglot Bible* printed in 1569–72 for Philip II of Spain which he presented to the emperor, but which was apparently not very greatly admired, being returned to the Jesuit father in 1595. Other paintings presented to the emperor included a copy of the *Byzantine Virgin* in the Borghese Chapel of the Church of Santa Maria Maggiore in Rome and a picture of Christ which seem to have been received with greater pleasure.[9] It is also recorded that Father Monserrate, another member of the First Mission, showed Akbar sacred illustrated books during the journey to Kabul in 1581, so that the presence of European works of art at the Mughal court from at least 1580 onwards is well attested.[10]

The court at Fatehpur-Sikri was also a centre of much intellectual activity. The great theological and philosophical debates of the Ibādat-khānā were organized there, at first confined to Muslims, but later extended to include adherents of other religions as well (ca. 1575–80). The learned from all parts of the empire congregated here, among them the great Faiẓī, who came to court as early as 1567, followed by his father, the learned Shaikh Mubārak (1573), and his brother, the extraordinary Abū al-Faẓl, besides the great leaders of the Hindu, Jaina, and Zoroastrian faiths.

7. Two other buildings at Fatehpur-Sikri are dated, namely the so-called Bīrbal's House built in V. S. 1629/1572 and the Jāmaʿ-Masjid in A. H. 979/1571–2. See E. W. Smith, II, p. 13 and IV, p. 4. The precise dates of the Khwābgāh and Mariam's House, and thus of the paintings they contain are not known. The paintings are in a poor state of preservation; for reproductions done manually see *ibid.* I, Plates XV and CIX. Mariam's House seems to have received its name from local guides, well-known for the fertility of their imagination. They invented a Christian wife for Akbar, and declared the building to be her residence in fond hopes of getting extra rewards from foreign tourists. They buttressed their arguments by drawing attention to a very indistinct painting supposedly showing a fairy holding a child and interpreted it as a representation of the Annunciation. This may probably explain why some people, including a few scholars, associate the several fairies painted on its walls with European influence while they are clearly Mughal renderings after the Safawi manner. See *ibid.*, I, p. 31.
8. *Akbar Nāmā* III, pp. 207, 322. The mission seems to have left in 1575 and returned in December 1577.
9. See Maclagan, pp. 225–28 and V. Smith, p. 125.
10. European works of art, however, may have reached the court earlier, either with the return of Ḥājī Ḥabībullāh's mission from Goa in 1577, on the occasion of Akbar's first meeting with the Portuguese in 1573, or even earlier.

Akbar himself was a man of deep spiritual sensitivities, and in 1578 underwent yet another profound mystical experience. His disenchantment with organised religion continued to increase. In June, 1579, he read the *khuṭba* personally in the Jāmaᶜ-masjid and in September of the same year, he was endowed with powers which made him the final adjudicator of any differences of opinion with regard to the interpretation of Muslim law, provided these were in conformity with the Qurᶜān Sharīf and for the benefit of the nation. The spirit of tolerance evidenced by him as early as 1562 culminated in the promulgation of the Dīn-i Ilāhī, or Tauḥīd-i Ilāhī, in 1582, considered to be a religion by some scholars, though this is denied by others.[11] *Ṣulḥ-i kul,* Peace with all, was the basic belief, and universal tolerance and acceptance of what was considered to be the best in all religions under the spiritual leadership of the emperor was enjoined upon members.

Akbar set out for the Punjab, leaving Fatehpur-Sikri in August 1585, and arrived at Lahore in 1586 where the court was in residence for the next twelve years. The empire continued to flourish: Kabul was absorbed (1585), Kashmir annexed (1586), Sind conquered (1591), followed by the subjugation of Baluchistan and Kandahar in 1595. The Deccan was invaded, but, Mughal troops being stymied, Akbar took the field in person, leaving Lahore in November 1598. After a short halt at Agra, he proceeded towards Khandesh around the middle of 1599, the great event of the campaign being the capture of the impregnable fort of Asirgarh in January 1601. Akbar returned to Agra shortly thereafter, his last years saddened by the misbehaviour of his son Prince Salīm, the murder of Abū al-Faẓl and the death of another son, Dāniyāl. He himself was claimed by death in November 1605 after one of the most glorious and memorable reigns India has ever seen.

Akbar as a patron of the arts

Akbar's reign is also remarkable for extensive and innovative activity in the visual arts, and, considering the nature of the times, it would have been unusual if this had not been the case. There was a striking revival of architecture in which non-Islamic styles, once again given full play, were successfully assimilated and reinterpreted in a great outburst of creative activity, as is clearly seen in the buildings of Agra Fort and Fatehpur-Sikri. Akbar, a person of wide interests, seems to have taken a personal part in this revival, himself supervising to some extent these colossal undertakings, as

11. For a brief discussion see Srivastava I, pp. 310 ff. We ourselves are inclined to believe that it was not a new religion, but partook more of the nature of a select order. Monserrate's account has to be treated with caution as he was a disillusioned man from failure to convert the emperor to Christianity, and being a foreigner relying on an interpreter, quite capable of misunderstanding the subtleties of the situation.

demonstrated by miniatures from the royal copy of the *Akbar-nāma* (Victoria and Albert Museum Nos. 2–1896/45, 46, 91) and by the character of the architecture itself, which shows startling departures from preceding traditions. Although literary evidence with regard to the part played by Akbar in the development of architecture is scarce, we have much more information regarding his patronage of painting, an art which we know Akbar to have himself practiced in his youth. No paintings from his own hand are known to have survived, so we do not know the extent of his personal skill in the art, but it was surely enough for him to play a positive part in the formulation of the style developing in his atelier. This style represents an even more radical shift than that of the architecture, and is difficult to understand without giving consideration to its patron's distinctive concepts of what was desirable and his ability to make the style conform to his ideals by his knowledge and connoisseurship.

The thirty-fourth *āʾīn* of the *Āʾīn-i Akbarī*, entitled the *"Āʾīn-i Taṣwīr-khāna,"* gives us valuable information on Akbar's attitude towards painting, the imperial atelier and the manner of its working.[12] The author states Akbar's predilection for the art from his earliest youth and his answer to those who disapproved of it. Far from being a sacrilegious attempt to imitate the works of God, the painter, however, life-like his work, knew that he was incapable of giving it life, and would be in a special position to learn that God alone is the giver of life. This idea would also be evoked in the person who saw the painting, and would thus become a source of wisdom.

That Akbar took a keen interest in the art resulted in its flourishing. Works of painters were inspected by him once a week, when rewards were distributed or salaries increased according to the excellence displayed.[13] This system, coupled with a patron well-versed in the art, must have encouraged the artists considerably in moulding their style to suit the patron's taste. It is also probable that ascriptions to painters found on the margins of many Mughal paintings of the period were primarily written down in order to facilitate this inspection. Further evidence of Akbar's influence upon the atelier is contained in a remark which attributes the excellence of Khwāja ʿAbd al-Ṣamad to the "transmuting glance of the emperor [which] has raised him to a more sublime level, and his images have gained in depth of spirit."[14] Nor can it be dismissed as conventional praise in the manner of courtly biographers. In this connection, it is interesting to recall the intimate association

12. The *Āʾīn* was the result of seven year's work by Abū al-Faẓl and was largely completed by 1597–8. See Appendix C for a new translation of the relevant section.
13. Cf. *Āʾīn* III, p. 347: "Artists of all kinds are employed at the imperial court where their work is subjected to the test of criticism."
14. Appendix C, p. 183.

between artist and patron in neighbouring Persia, where it was said of Shāh Ṭahmāsp, that so great was his competence that the great masters of his atelier forebore from putting the finishing touches to their work before submitting it for the Shāh's correction and approval.[15] The discovery of the painter Dasavanta by Akbar himself is another indication of the emperor's concern for unearthing and encouraging talent. His (Dasavanta's) being handed over to Khwāja ʿAbd al-Ṣamad for training indicates a system by which new artists or less accomplished ones were trained under the supervision of the well-established masters.[16]

Enough has been said of the part played by Akbar's taste in the evolution of the Mughal style; what this taste was can be easily gathered from the surviving examples dateable to the early years of his reign, especially the celebrated *Qiṣṣa-i Amīr Ḥamza (Ḥamza-nāma)* specifically and intimately associated with him,[17] and also the more preciously executed miniatures of the 1568 *ʿĀshiqa* and the 1570 *Anwār-i Suhailī* (Pls. 35–36; 39–42). Already in these we see a style in which "inanimate objects appear to come alive" and where "delicacy of work, clarity of line and boldness of execution" are much emphasised.[18] How this style was made possible, given the state of painting in India when Akbar ascended the throne, is revealed by the Cleveland *Ṭūṭī-nāma,* and will be discussed in some detail in due course.

Other information about Mughal painting which we can glean from the *Āʾīn* is that by 1597–8, its date of completion, the atelier was quite large with over a hundred masters and many more painters of good and middling quality. Perfection in the art, quite obviously, was not a prerequisite to gaining employment in the atelier, but had nevertheless been achieved there so that few nations in the world were able to display such glories. Four painters are singled out for special consideration: Mīr Sayyid ʿAlī, Khwāja ʿAbd al-Ṣamad, Dasavanta, and Basāvana, in that order. Dasavanta is described as having no peers, but Abū al-Fażl displays his own skillful judgment by gently advancing the claims of Basāvana "whom many perspicacious connoisseurs prefer to Dasavanta" by virtue of his excellence in *ṭarrāḥī* (designing), *chihra-kushāʾī* (portraiture), *rang-āmezī* (blending of colours, colouring),

15. Ṣādiqī Beg Afshār quoted by Welch, *King's Book of Kings,* p. 69.
16. Appendix C, p. 183.
17. Mīr ʿAlā al-Daula states that the book was commanded by Akbar himself, Appendix B, p. 283. Ḥājī Muḥammad ʿĀrif names the *Ḥamza-nāma* to exemplify Akbar's character as a designer of marvels, Appendix D, p. 187.
18. Appendix C, p. 183. A movement away from the decorative emphasis of Persian art to more naturalistic preferences can also be gleaned from other remarks of Akbar. Thus in connection with the presentation of *Nal-Daman* by Faiżī, Abū al-Fazl states that Akbar paid less regard to poetry for it was based on fancy and fiction, *Akbar Nāmā* III, p. 1015. The same reasons are given by Badāyūnī II, p. 329 for Akbar turning away from Islamic works to Hindu books like the *Mahābhārata.* Apparently by 1582, when the translation of the *Mahābhārata* was begun, if we are to trust Badāyūnī, everything that was Hindu was real and true in the emperor's estimation!

and *mānind-nigārī* (taking likeness). Traditions of critical connoisseurship observed in the writings of Mirzā Ḥaidar seem to have been flourishing in Akbar's court as well. Besides the above, thirteen other painters are singled out by name, Tārā being of some significance, as a painting in the Cleveland *Ṭūṭī-nāma* is ascribed to him (fol. 60r).

Ascriptions to painters in Akbar period painting

A large proportion of the miniatures of the Akbar period now known to us carry short labels written on the margins. These name the painter or painters responsible for the work. They were probably put there, as stated earlier, to facilitate imperial inspection, or for purposes of record, and were generally made by an officer of the atelier. Though they are not, for the most part, artists' signatures, they are, nevertheless, contemporary with the painting, and stylistic analysis has, in the main, confirmed their reliability. Sometimes miniatures do not seem to possess ascriptions, for manuscripts were often rebound, and if in this process the pages had to be recut, the ascriptions were trimmed away, though in some instances, the earlier ascriptions were recopied when this happened. It is also possible that on occasion ascriptions may have never been written, whatever the reason may be; and we also know that some ascriptions, written in later, are quite fanciful and untrustworthy. It is thus necessary to always take into account the style of miniatures when attempting to ascertain their painters; but, at the same time, unless there is some valid reason to the contrary, it is useful to proceed on the assumption that the ascriptions are correct, discarding this assumption if it is invalidated on grounds of style.

Ascriptions are obviously very helpful in determining the styles of individual painters, but they also yield other useful data. They confirm, for example, the existence of a large number of painters at the court, and that of these, the overwhelming majority were Indians. From an analysis of the names, it has been estimated that over three-fourths were Hindus, and the remainder Muslims, and of these Muslims, many were Indians,[19] an estimation that is fully in keeping with the style. That artists were drawn from several parts of the empire is confirmed, not only by types of names, but also by specific regional appellations used for some painters, such as *gujarātī* (from Gujarat), *kashmīrī* (from Kashmir), *gvāliyarī* (from Gwalior), *lāhorī* (from Lahore), etc. Sometimes the name of the father is also added on to that of the son (eg. Bihzād, son of ʿAbd al-Ṣamad; Manohara, son of Basāvana), from which one can estimate the hereditary character of the profession.

19. Arnold, *Beatty Catalogue* I, p. xxviii.

Particularly valuable are ascriptions which assign miniatures to a single person, for they make it possible to distinguish the styles of the individual painters. Ascriptions of this type often merely consist of the name of the painter, or his name preceded by the word *ʿamal,* in this context meaning "the work of."[20] There are some manuscripts in which almost all the miniatures are done by individual painters working unassisted, but there are others in which most miniatures are the result of joint effort. These last present much more complex stylistic problems, but careful and comprehensive analysis is most rewarding and gives us an intimate glimpse into the actual processes at work in the atelier, such as the interplay and struggle between two different concepts of a desirable style, the degree of interest taken by the master artist in a particular miniature, the extent of latitude given to the assistants and so on, so that the whole atmosphere of the atelier can come to life in a vivid way.

In jointly produced miniatures, the work was generally divided among two painters. An ascription reading *ṭarḥ basāwan ʿamal pāras* thus indicating that Basāvana was responsible for the design, the colouring being done by Pārasa.[21] Sometimes paintings done by the same two artists can be visually quite different, and this is often to be accounted for by the degree to which each artist was involved. A master designer may thus rework his colouring assistant's work until it looks more like his own hand *in toto,* or he may do but little, even allowing his design to be obscured by the style of the colourist, or each may just do enough to quite confuse the student of their work. An analysis of the paintings and the ascriptions also reveals that it is the masters that generally did the designing, and that the colourists were quite often second echelon artists. There are instances of painters beginning as colourists and becoming designers (Miskīn, for example, is a colourist in the Jaipur *Razm-nāma,* but is a designer in the Victoria and Albert *Akbar-nāma);* and in truly ambitious manuscripts, master artists who are generally designers do the work of colourists (eg. in the Jaipur *Razm-nāma).* In addition to these two, we find another category of work, namely the painting of portraits, and sometimes this task is assigned to a third artist, different from both the designer and the colourist. There are instances of paintings in which the designer has also done the portraiture, but none, so far as we are aware, where the colourist attempts this difficult undertaking.

20. The same word when used in the context of joint work signifies the colourist.

21. This is not to be understood narrowly in the sense of sketch or outline being done by Basāvana, the colours being filled in by Pārasa but more broadly, Basāvana being the designer of the painting, its master artist, while Pārasa is his assistant. Also see Appendix C, p. 183, where colouring is referred to as *rang-āmezī.*

The atelier and its organization

Of the actual organization of the atelier, we know little, but the special treatment accorded to the *taswīr-khāna* by Abū al-Faẓl, who devotes an entire *āʾīn* to it, attests to its importance.[22] The atelier included both calligraphers and painters, and, theoretically at least, the art of calligraphy was given precedence. There were also a number of assistants, such as illuminators, gilders, margin makers, bookbinders, and presumably others performing manual tasks involved in the preparation of materials. Appointments to the *taswīr-khāna,* as in all state service, were made in military terms.

If Father Monserrate of the first Jesuit Mission is correct, the building which housed the painters as well as the gold-workers, weavers, and armourers was situated, at least between 1580 and 1582, close to the emperor's own residence at Fatephur-Sikri;[23] and considering the emperor's interest, this may well have been possible, though none of the surviving buildings seem large enough to have accommodated such an establishment. Perhaps only a few selected artists and craftsmen worked near the palace, the rest being housed elsewhere. The atelier, or at least part of it, seems to have followed the emperor, especially when he shifted his residence for any length of time. The colophons of manuscripts make this clear, for some of them state that they were made at Fatehpur-Sikri (Royal Asiatic Society *Gulistān* dated 1582), and others at Lahore (Fogg *Dīwān* of Anwarī dated 1588); nor could this be otherwise considering the keen interest taken by the emperor in the work of his painters. It is equally possible that some artists accompanied him on short expeditions as well, for the imperial camp was elaborately organised and the painters were hardly burdened with too much equipment.[24]

Varieties of work in early Akbar period painting

A general consideration of illustrated manuscripts of the Akbar period and the artists to whom the miniatures are ascribed reveals three clearly differentiated types

22. It is interesting to note that the title of the section is *Āʾīn-i taṣwīr-khāna,* not ***kitāb-khāna*** or ***kutub-khāna*** as one may have expected. The *taṣwīr-khāna* thus has an importance of its own and is not just considered as an appendage of the library. Cf. Zakhoder's introduction to *Calligraphers and Painters,* p. 3, fn. 20: "In medieval works the term ***kitāb-khāna*** includes both the library itself and the workshop in which the work of restoring and producing manuscripts decorated with painting was carried on."

23. *Mongolicae Legationis Commentarius* III, p. 643 quoted in Arnold, *Beatty Catalogue* I, p. xxii.

24. V. Smith's suggested identification of three persons accompanying Akbar on his lightning dash to Gujarat in August 1573, namely Jagannāth, Sānwal Dās and Tārā Chand Khwāṣ as three artists with similar names, however, is highly conjectural. The names were hardly uncommon among Hindus; and Jagannāth, I feel sure, is the same officer who was in charge of the van during the Mewar campaign while Sānwal Dās is elsewhere referred to as Gopāl Jādun's nephew and a personal attendant of the emperor. See Smith, pp. 84–85, fn. 3; *Akbar Nāmā* III, pp. 68–69, 244, and 651.

of work. The first consists of books with large numbers of fairly elaborate illustrations and in which—perhaps for this very reason—each miniature was assigned to generally two, and sometimes three artists, any other arrangement being considered impractical for the efficient completion of these extensive projects. The *Ḥamza-nāma* is the extreme example of this type, and also the earliest, the 1400 paintings, of exceptionally large size, making joint work inevitable if the project was to be completed in any reasonable length of time. As it were, the project is reported to have taken fifteen years to accomplish. The Jaipur *Razm-nāma,* with 169 miniatures, is on a relatively more modest scale, but still large enough, its various miniatures, with few exceptions, also being the result of joint effort. Work of this type was of varying quality, the level of achievement depending on the quality of the painters involved and the degree to which the collaboration was successful. The overall result was a generally high standard, the good artists covering up the deficiencies of those less talented, though by the same reckoning, the type of perfection achieved by the really great artists when left entirely to themselves was but seldom achieved. Miniatures in these large-scale manuscripts tend to be bold and straightforward in execution; and, with some exceptions, little attempt seems to have been made to strive for the high refinement and exquisite finish of what has been called the de luxe type of manuscript in which each miniature was almost always completed by a single artist. The Patna *Tārīkh* and the Victoria and Albert *Akbar-nāma* are among the other examples of what may be conveniently called manuscripts with jointly produced miniatures.

The de luxe type of manuscript possesses many fewer miniatures, hardly ever more than forty-five, and generally less. Every miniature is the work of a single master who lavished all the skill of his art in its execution, enriching and refining it to an extraordinary degree. Work of this type is found fairly early in the reign, the 1568 *ʿA<u>sh</u>iqa* (Pls. 35–36) and the 1570 *Anwār-i Suhailī* (Pls. 39–42),[25] with two and twenty-seven miniatures respectively, being painted (as we will show later) when the *Ḥamza-nāma* was actually in progress. The workmanship is very refined, and if these manuscripts had not been dated, the tendency would have been to assign them to a later period on the grounds of their "advanced" style. This demonstrates the importance of understanding the class of manuscript or miniature before we can assign it a correct place in the development of Mughal painting. Several other

25. The *ʿĀ<u>sh</u>iqa* of Amīr <u>Kh</u>usrau, also known as *Dewal Dewī <u>Kh</u>iẓr <u>Kh</u>ān* is in the collection of the National Museum, New Delhi, and is listed in *Manuscripts from Indian Collections,* pp. 96–97. It is dated Muharram 976/ June–July 1568 and not 1567 as is incorrectly stated above. The name of the scribe is given as Sulṭān Bāyazīd b. Mīr Niẓām. The *Anwār-i Suhailī* is in the library of the School of Oriental and African Studies, London. It is dated 22 Rabīʿ II 978/23 September 1570, the name of the scribe being Muḥibullāh b. Ḥasan. See *Art of India and Pakistan,* p. 142, Cat. No. 636. The number of miniatures is more than the twenty-four given by Gray. My notes indicate that they are twenty-seven in number.

manuscripts of this de luxe class are known, and include a *Dīwān* of Anwarī dated 1588, a *Khamsa* of Niẓāmī dated 1595, a *Bahāristān* of Jāmī dated 1595, and a *Khamsa* of Amīr Khusrau dated 1597–8 in the Walters Art Gallery.[26]

In addition to these two types of manuscripts, yet a third type can be posited. It contains paintings which are not the products of joint effort, but once again of individual painters working unassisted, except that the work is not of a rich and luxurious character as in the de luxe category, and the artists are not all masters. The compositions are also not very elaborate, and some miniatures may cover only half the page or an even smaller area. The total number of paintings can often be quite large. The style of the miniatures is comparable to those produced jointly, and is on the whole bold rather than delicate, the artists giving the impression of working rapidly and freely. This is not to say that paintings in this kind of manuscript are of a generally lesser quality, for a great master's excellence will almost invariably stand out, and the quick and spontaneous rendering often makes up in power what is lacking in finesse.[27] Because of the large number of painters employed, one can not only see clearly the style of each individual artist, the quality of their work not being obscured by joint production, but we can also get a very good idea of the broad range of achievement of the school as a whole at the time when any given manuscript of this type was produced. As in the case of de luxe manuscripts, this type of work is to be seen as early as the *Ḥamza-nāma* period in a fragmentary and remounted manuscript dealing with astrology, magic spells, and talismans in the Raza Library, Rampur (Pls. 43–44).[28] The work has the breadth and fluency of the *Ḥamza-nāma,* but does not possess the refined and exquisite finish of de luxe manuscripts like the 1568 *ʿĀshiqa* or the 1570 *Anwār-ī Suhailī* (cf. Pl. 43 and Pls. 35, 39), especially in the small miniatures, though the seven large paintings illustrating the zodiac are fairly accomplished. The British Museum *Dārāb-nāma* with 149 miniatures (Pls. 45–46) is also a manuscript of this type,[29] and we see here the entire gamut of Mughal painting ranging from an exquisite masterpiece by Basāvana (Welch, "Basawan," Fig. 1) to the fairly primitive and archaic efforts of the Lahore painters (Pl. 46). Reliable ascriptions to forty-five painters (forty-four painters, if we exclude Khwāja ʿAbd al-Ṣamad, who is recorded as having merely improved a

26. Welch, *Art of Mughal India,* p. 162, Cat. No. 4; Welch, "Akbar's Khamsa," pp. 87–96 and Martin, p. 81, Pls. 178–81; and Grube, *Classical Style,* pp. 202–03, Cat. No. 91.

27. Sometimes painters also contributed works of de luxe quality to this class of manuscript. The painting by Basāvana in the British Museum *Dārāb-nāma,* for example, is as sumptuous as his miniature in the Bodleian *Bahāristān* which is a manuscript of the de luxe class. Rf. Welch, "Basawan," Figs. 1 and 3.

28. See Khandalavala and Mittal, "MS. of Tilasm and Zodiac," pp. 9–20. The manuscript seems to have been illustrated by a more select group of painters than are to be seen, for example, in the British Museum *Dārāb-nāma.*

29. Rieu, *Catalogue,* p. 241. The manuscript is unpublished except for stray miniatures.

painting by Bihzād) are to be found, all of them working by themselves except in a very few instances. Fifteen of these have contributed four or more miniatures, the maximum by Bhagavāna who is ascribed nine, the rest painting from one to three. The manuscript is thus in its scope and arrangement, as we will shortly see, very reminiscent of the Cleveland *Ṭūṭī-nāma,* which has far fewer ascriptions, but had, as far as one can judge on grounds of style, a comparable number of artists working on its 218 miniatures, their range being even more diverse, from masterpieces by Basāvana and Dasavanta at one end, to artists who had barely stepped out of the pre-Akbar idioms on the other. Among other manuscripts of this type, consisting of singly painted miniatures, are an *Anwār-i Suhailī,* unfortunately in a very fragmentary condition due to the ravages of fire, recently gifted to the Prince of Wales Museum of Western India, and the *Ṭūṭī-nāma* in the A. Chester Beatty Library in Dublin (Pls. 47–61)[30]. Unfortunately neither of these now possesses the ascriptions which they once presumably carried. It is also interesting to note how the wide divergence of styles among the various artists working on early manuscripts of this type is lessened as the style itself gets well established in the course of time and the painters become increasingly comfortable in it.[31] It is only towards the close of Akbar's reign, when a whole new generation of artists is trained and ready, that manuscripts of this type, with singly painted miniatures, begin to acquire a degree of true homogeneity and an overall richness and refinement hardly to be distinguished from that of the de luxe manuscripts. The necessity of joint work was also at this stage no longer felt, for there was now no shortage of trained painters, and all Mughal painting, within the limits of individual competence, came to possess a fairly uniform quality. The A. Chester Beatty *Akbar-nāma,* begun towards the close of Akbar's reign and probably completed in the next, and the National Museum *Bābar-nāma* are good examples of this new phase of Mughal painting.[32] With Jahāngīr's accession, the transformation to a consistent, homogenous style was complete, personal taste as well as the history of the art resulting in works of uniform richness and excellence. From this time on, it even becomes possible to consider attributing works to sub-imperial patronage or to another period altogether on grounds of quality alone. The great variety of work in the Akbar period, particularly in the earlier part of the reign, makes the application of this criterion

30. The Beatty manuscript has 113 illustrations, but several remounted folios which once belonged to it are scattered in collections throughout the world. See *Art of India and Pakistan,* p. 142, Cat. No. 637, Pl. 120, and Binney, *Catalogue Mughal and Deccani Schools,* p. 33, Cat. No. 14.

31. Archaistic memories, however, survive as late as the 1590s and are present even in a royal manuscript of such excellence as the Victoria and Albert *Akbar-nāma* (Pl. 63).

32. The first volume of the *Akbar-nāma,* now in the British Museum, has a miniature dated 21 Shaʿbān 1012/ 25 January 1604. Arnold, *Beatty Catalogue* I, pp. 4–12, and II, Pls. 6–37; *Manuscripts from Indian Collections,* p. 106.

exceedingly difficult. True, there did exist what I have called a popular Mughal style in the reign of Akbar, but examples of these works possess a marked internal stylistic consistency, in contrast to the great variety of Mughal work in its early formative stages.[33]

The development of Mughal painting in the early Akbar period: the Ḥamza-nāma

Bearing in mind the stylistic varieties of early Akbar period painting, we can now attempt to correctly assess the half-a-dozen or so surviving manuscripts of the early Akbar school. This together with our knowledge of the pre-Mughal styles as outlined in Chapter II will have adequately prepared us for a detailed consideration of the Cleveland *Ṭūṭī-nāma,* which can then follow.

The most important manuscript of this early period, and perhaps of all Akbari painting, is the *Qiṣṣa-i Amīr Ḥamza,* to call it by the name consistently given to the manuscript in contemporary literature (Pls. 1–34).[34] It consists of legendary tales of high adventure in Ceylon, China, Central Asia and Rūm, centering around Ḥamza who was originally a Persian insurrectionary leader from Sistan during the reign of the Caliph Hārūn al-Rashīd, but was later identified with the uncle of the Prophet Muḥammad who had the same name.[35] The tales, which are of Persian origin, were also very popular in India, and were a particular favourite with the emperor Akbar. According to the *Ma᾿āsir al-Umarā᾿*, so fond was he of these stories that he used to recite them in the harem like a story teller;[36] and the *Akbar-nāma* relates how he relaxed listening to these stories after a strenuous elephant hunt near Narwar in 1564.[37] The emperor ordered a new version to be prepared in the then current Persian language, and had it most sumptuously scribed and illustrated on a magnificent and grand scale. The colossal project was completed in fourteen volumes

33. See P. Chandra, "Ustad Salivahana," pp. 25–46.
34. For a comprehensive bibliography see Grube, *Islamic Paintings,* pp. 254–55, fn. 1, the most important work being the magnificent publication by Glück. The account given here, in so far as literary testimony is concerned, is based upon relevant texts as translated in Appendices B–D, where the manuscript is always called the *Qiṣṣā-i Amīr Ḥamza.*
35. Meredith-Owens, "Ḥamza," pp. 152–54.
36. I, p. 454. The work, a scholarly compilation of biographies of Mughal noblemen was probably written between 1741–47 with some later additions (Preface, p. v). The author often used earlier sources which are now lost, and his reliability depends upon the sources he used. The account of the *Ḥamza-nāma* is largely abstracted from Mīr ᶜAlā al-Daula (Appendix B), though he gives some details not found elsewhere.
37. *Akbar Nāmā* II, p. 343.

and had a total of 1400 paintings.[38] Each of these paintings was of exceptionally large size, approximately 67 by 51 cms., painted in cloth, the text placed on the reverse, and may have had borders decorated with floral patterns.[39] The work evoked universal admiration among contemporaries, in the eighteenth century, and does so to the present day. Sometime after the mid-eighteenth century when it was in the Imperial Library, having survived Nādir Sh̲āh's sack of Delhi (1739),[40] it left that library, and some paintings seem to have fallen into the hands of what Akbar would have called a "bigoted follower of the Law" who erased the faces of human beings, animals, and birds, some of these erasures being subsequently repainted by an artist who tried to make amends for this barbarous act. As it is, of the 1400 paintings, about 140, or one-tenth of the total, seems to have survived,[41] but these are enough to give us an idea of what the original manuscript must have been like. "No one has seen such another gem nor was there anything equal to it in the establishment of any king."[42]

The important place which the *Ḥamza-nāma* occupies in the history of Mughal painting has been recognized by all; but its date, though the subject of much

38. Abū al-Faẓl, the official court chronicler writing in the *Āʾīn* which was completed in 1597–8, well after the *Ḥamza-nāma* had been finished, gives 12 volumes and 1400 paintings (Appendix C, p. 184). Mīr ʿAlā al-Daula writing earlier and before the *Ḥamza-nāma* had been completed (See Appendix B, p. 180) states that it was planned in 12 volumes, each volume possessing a hundred paintings, giving a projected total of 1200. Badāyūnī, who began his *Muntak̲h̲ab al-Tawārīk̲h̲* in 1590 (*Encyclopaedia of Islam,* p. 857) did not have the resources of the state at his disposal as did Abū al-Faẓl, and it is not surprising to note that his "chronology is less precise than that of the *Akbar-nāma*" (V. Smith, p. 339) as are some of his facts. He thus states at one place that the *Ḥamza-nāma* consisted of 17 volumes (Badāyūnī II, p. 329) and in another that it consisted of 16 (*ibid.,* III, p. 292) indicating that he was not sure of the exact number. The *Maʾās̲ir al-Umarāʾ*, though a late work written ca. 1741–7, apparently following Mīr ʿAlā al-Daula, also states that the *Ḥamza-nāma* was in 12 volumes, each volume of 100 folios, but each folio, according to it had two paintings which is manifestly incorrect, p. 454. It also states that 50 painters were engaged upon it, though the Munich copy of the *Nafāʾis* gives the number as 30. Muḥammad ʿĀrif's *Tārīk̲h̲* written between 1581–4 (Appendix D, p. 187) states that there were as many as 100 painters, decorators, and binders engaged on the manuscript. He does not give us the number of volumes or paintings but states that every two years one volume was prepared, agreeing fairly closely with Mīr ʿAlā al-Daula's account of four volumes being prepared in seven years. Glück, on the basis of a painstaking reconstruction was able to state that the manuscript was indeed in 14 volumes, with 100 paintings per volume, making a total of 1400, Abū al-Faẓl being thus correct as far as the total number of pictures was concerned but in error regarding the number of volumes (Glück, p. 155).
39. Both Mīr ʿAlā al-Daula, Appendix B. p. 180, and Muhammad ʿĀrif, Appendix D. p. 187, clearly imply that the pictures were square in shape, while this is not the case in surviving examples. Perhaps the measurements given by the authors included the margins which were altered in the course of time. Muḥammad ʿĀrif indeed states that the margins were decorated with floral patterns which are no longer present (Appendix D, p. 187), and signs of later repairs are obvious. He also refers to the cloth carrier.
40. This is to be inferred from the *Maāt̲h̲ir-ul-Umarā* I, p. 455, where the author speaks of it as being present in the imperial library.
41. For a list of most of the miniatures presently known see Grube, *Islamic Paintings,* p. 256, fn. 4.
42. *Maāt̲h̲ir-ul-Umarā* I, p. 455.

speculation, has remained uncertain, even if it is universally regarded as the earliest work of the Mughal style. The reasons for this are not far to seek, it being difficult to come to a judgment on the basis of a small fraction of a huge manuscript that seems to have taken a period of about fifteen years to complete. Because it is the first work of the Mughal school, there are no earlier paintings available for comparison, nor are related Persian miniatures of much help, for the style of the *Ḥamza-nāma* has moved considerably from them. And all that a comparison with subsequent Akbar period painting does is to establish the priority of the *Ḥamza-nāma,* but this is not saying much. Nor has it been possible to set up any internal sequence for the paintings themselves. Evolutionary hypotheses postulating a development from the "archaic" to the "refined" or from the proto-Persian to the fully Mughal do not help, for painters with widely varying talents seem to be simultaneously at work, a possibility which would be but natural at this formative stage; and the fact that the paintings of the *Ḥamza-nāma* were jointly executed, obscuring the style of individual painters, further adds to our difficulties. Glück suggested a development consistent with the sequence of the story, but this is true in only a very broad and general way. It could thus be asserted that the first four volumes or so, painted, as we now know, under the supervision of Mīr Sayyid ʿAlī, are earlier than the last ten which were painted under Khwāja ʿAbd al-Ṣamad; but here also we run into difficulties, for the text must have taken a shorter time to complete than the paintings, and later events could have been painted earlier or vice versa. Actually, the paintings are on cloth and the text is written on paper, a feature which would additionally suggest the independence of the paintings from the text, each being done separately and subsequently brought together.

In the above circumstances, when we have a group of paintings of such a baffling character, one has perforce to rely on written testimony, circumstantial or otherwise, if any is available. Materials of this type are notoriously rare for Indian art, but, fortunately, the *Ḥamza-nāma,* being such a spectacular production, did attract comment from contemporary chroniclers. These too, however, were either not precise enough to be satisfactory, or have been misunderstood, accounting for the wide divergence of opinion in previous considerations of the subject. Many scholars, including Glück, Stanley Clarke, Stchoukine and Gray, considered work on the huge undertaking to have begun somewhere around the time of Humāyūn's return to India in 1555, with the major part of the manuscript being completed, broadly speaking, sometime during the first half of the reign of Akbar. The relative proximity of the *Ḥamza-nāma* to the Safawi style as practiced at the court of Shāh Ṭahmāsp encouraged the belief that it was begun by Humāyūn, but after the Persian exile and late in his reign, probably around 1550–55, a little before his return to the imperial throne in India; the terminal date, however, often rested on somewhat extraneous considerations. Gray, for example, found it difficult to believe that work on the *Ḥamza-nāma* could have continued after 1579. The emperor, according to him, began to profess markedly liberal religious views after that time, and it was inconceivable to think of him extending patronage to such an emphatically sectarian work like the *Ḥamza-nāma* after that date.[43] Rai Krishnadasa was among the first to assert an exclusively Akbari origin for the *Ḥamza-nāma,* relying upon the testimony of Badāyūnī and Abū al-Faẓl, a position he maintains in later publications as well.[44] The recent tendency has been to accept the *Ḥamza-nāma* as a work of the Akbar period, though an origin in the reign of Humāyūn continues to hold some support.[45]

Some rather important and reliable evidence bearing upon this problem is provided by the testimony of the *Nafā'is al-Ma'āsir* of Mīr ʿAlā al-Daula Qazwīnī, a work of the Akbar period. Chaghatai and Khandalavala[46] attempted to utilize it, but, as their efforts are of an uncritical nature, notably those of Khandalavala which are full of distortions and misrepresentations as well, the matter deserves to be re-examined thoroughly, more so because of the close relationship between the Cleveland *Ṭūṭī-nāma* and the *Ḥamza-nāma.*[47]

According to Mīr ʿAlā al-Daula, the *Ḥamza-nāma* was commissioned by Akbar and by none other. It was planned to be completed in twelve volumes with 1200 paintings, and was initially under the charge of Mīr Sayyid ʿAlī who worked on it for seven years completing four volumes. Sometime after the four volumes had been completed, Mīr Sayyid ʿAlī left on a pilgrimage for Mecca and the work was entrusted to Khwāja ʿAbd al-Ṣamad who proceeded on it with some speed and a notable reduction of expenditure. The section of the *Nafā'is al-Ma'āsir* in which the reference to Mīr Sayyid ʿAlī and the *Ḥamza-nāma* occurs was written sometime between A. H. 973 and A. H. 979, so that one can clearly conclude that the *Ḥamza-nāma* was begun between A. H. 966 and A. H. 972 (see Table 1).

But when was the *Ḥamza-nāma* finished? Mīr ʿAlā al-Daula is of no help in answering this question; but we are aided by Badāyūnī who says that the work took fifteen years to complete.[48] Assuming him to be correct, we can affirm, on the basis

43. Gray, *Painting of India,* p. 78.
44. *Bhārata kī citra-kalā,* pp. 124 ff. and *Mughal Miniatures,* Pl. I.
45. Grube, *Islamic Paintings,* p. 252, relying on Chaghatai, "Mir Sayyid Ali," p. 26, continues to believe in a Humāyūn period origin for the *Ḥamza-nāma.*
46. *Ibid.,* and Khandalavala and Mittal, "MS. of Tilasm and Zodiac," pp. 12-13.
47. For an account of Mīr ʿAlā al-Daula, his work, its date with particular reference to the account of Mīr Sayyid ʿAlī and a translation of the relevant section see Appendix B, the discussion above being based upon the evidence which has been presented there in some detail. Table 1 has been prepared to assist the reader in following the arguments presented.
48. Badāyūnī II, p. 329. The statement seems to be reasonable if we remember Khwāja ʿAbd al-Ṣamad's great endeavours to bring the work to completion after the comparatively slow start by Mīr Sayyid ʿAlī. See Appendix B, p. 181.

Table 1

Chronology of the *Ḥamza-nāma*

A. H. 966	*Ḥamza-nāma* begun after this date (ʿAlā al-Daula)
7	
8	
9	
970	OUR PROBABLE DATE FOR INCEPTION OF *ḤAMZA-NĀMA*
1	
2	Last date for inception of *Ḥamza-nāma* (ʿAlā al-Daula) First 4 volumes of *Ḥamza-nāma* completed after this date (ʿAlā al-Daula)
3	ʿAlā al-Daula begins *Nafāʾis al-Maʾāṣir*
4	
5	
6	Date of *ʿĀshiqa,* National Museum
7	*Ḥamza-nāma* definitely in progress between these dates (ʿAlā al-Daula and Badāyūnī) [973–980]
8	Last date for completion of first 4 volumes of *Ḥamza-nāma* (ʿAlā al-Daula) Date of *Anwār-i Suhailī,* London School of Oriental Studies
9	ʿAlā al-Daula completes *Nafāʾis al-Maʾāṣir*
980	ʿAbd al-Ṣamad begins superintending *Ḥamza-nāma* after Sayyid ʿAlī's departure for Mecca (ʿAlā al-Daula) [980–981]
1	
2	
3	
4	
5	OUR PROBABLE DATE FOR COMPLETION OF *ḤAMZA-NĀMA*
6	ʿAbd al-Ṣamad appointed to mint (Abū al-Faẓl)
7	Last date for completion of *Ḥamza-nāma* (ʿAlā al-Daula and Badāyūnī)

of his testimony and that of Mīr ʿAlā al-Daula, that theoretically the earliest date for the completion of the *Ḥamza-nāma* would be A. H. 980, and the latest date would be A. H. 987.[49] It is also clear from this that whatever may be the exact dates for the beginning and end of the *Ḥamza-nāma,* it was certainly being painted between A. H. 973 and 980.

49. It is necessary to consider here the authoritatively stated but entirely unwarranted views of Khandalavala presented in a joint article with Mittal, "MS. of Tilasm and Zodiac," pp. 11-13. According to him there should never have been any doubts about the date of the *Ḥamza-nāma* for Badāyūnī clearly "indicates that it was *completed in 1582* (italics mine) and had taken fifteen years to complete, which means that it was in production from A. D. 1567 to A. D. 1582." Now Badāyūnī says nothing of the sort, as even a quick reading of the English translation or the Persian text clearly reveals. Recounting one of the notable events of the year A. H. 990/1582, he refers to the translation of the *Mahābhārata* at Akbar's behest, and feels that this was due to a change in the emperor's taste. "When he had (*nawīsaʾī dand,* in the sense of 'when he had

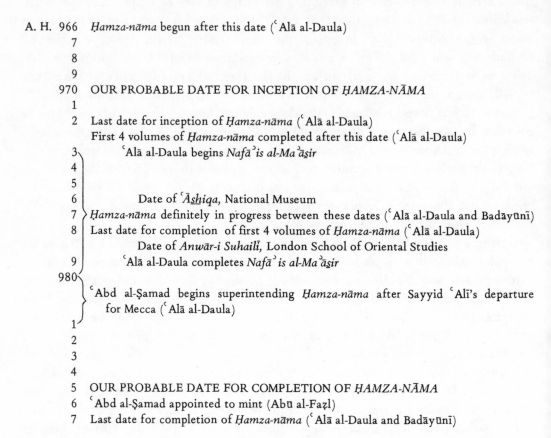

The question now is whether it is possible to further narrow this broad range of dates. The evidence provided by Mīr ʿAlā al-Daula is again of some help. According to the Munich manuscript, Khwāja ʿAbd al-Ṣamad was at work on the *Ḥamza-nāma* in A. H. 980–81, with Mīr Sayyid ʿAlī's departure for Mecca. Though exact dates are not given, there is the clear implication that a considerable amount of work remained to be done, perhaps as many as ten volumes, the *Ḥamza-nāma,* as we know, being extended to fourteen volumes. We would therefore suggest that the *Ḥamza-nāma* was, in all likelihood, completed around A. H. 985, following which in A. H. 986, Khwāja ʿAbd al-Ṣamad was appointed to become the master of the mint at the capital of Fatehpur-Sikri in connection with the monetary reforms instituted by Akbar. If this date is correct, the *Ḥamza-nāma* was begun fifteen years earlier i. e. in A. H. 970. An earlier date for the completion of the *Ḥamza-nāma* does not appear likely as it would give Khwāja ʿAbd al-Ṣamad less than four to five years to complete the work left unfinished by Mīr Sayyid ʿAlī, which would seem to be too short a period even if he expedited the work considerably. If we were to consider a later date for the completion of the *Ḥamza-nāma,* it could in no case be after A. H. 987, as shown earlier. It would thus appear to us, all things considered, that the most likely dates for the *Ḥamza-nāma* are ca. A. H. 970–85/A. D. 1562–77.

The above date has the advantage of not conflicting with any of the testimony including conceivably that of Muḥammad ʿĀrif and even perhaps the *Akbar-nāma.*[50]

had') the *Shāh-nāma* and the *Story of Amīr Ḥamza* in seventeen volumes transcribed in fifteen years, . . . it suddenly came into his mind that these books were nothing but poetry and fiction (Badāyūnī II, p. 329; Text II, p. 20)." Nowhere does Badāyūnī, or the English translation, state that the *Ḥamza-nāma* was completed in 1582; all that is said is that the book had taken fifteen years to make and had been already completed sometime before 1582, which could mean any fifteen years during the reign of Akbar before 1582. Khandalavala next proceeds to confirm his dates by an even more amazing distortion of evidence contained in the *Nafāʾis al-Maʾāsir.* According to him the date of the last event mentioned in the book was A. H. 982/ 1574, and since the book also stated that seven years had passed since work on the *Ḥamza-nāma* had begun, it was commenced in 1567. This conclusion, needless to say, could only be correct if the *Nafāʾis al-Māʾaṣir* stated that the *Ḥamza-nāma* had been in progress for seven years in the context of events occurring in 1574. And for this Khandalavala has produced no evidence, nor is there any in the manuscripts we have examined, including the Rampur copy which seems to form the basis of his remarks. He also notes that the date of 1567 for the inception of the *Ḥamza-nāma* arrived at by him "tallies exactly with Badāyūnī's account." We have seen on what dubious and flimsy grounds both these dates, the one supposedly tallying with the other have been arrived at. What Khandalavala seems to have really done is to arbitrarily select a date in the *Nafāʾis al-Maʾāsir,* the so-called date of the last event (1574) for no other reason but to wrongly demonstrate a coincidence between the evidence of that work and Badāyūnī as far as the date of the inception of the *Ḥamza-nāma* is concerned. Thus his conclusions and even more his methods, according to which the *Ḥamza-nāma* was begun in 1567 and completed in 1582 lack all credence and have to be entirely discarded.

50. See Appendix D, p. 186. The *Akbar Nāmā* II, pp. 343–44 reveals Akbar listening "for the sake of delight and pleasure" to Darbār Khān reciting the story of Amīr Ḥamza after an elephant hunt sometime in July-August 1564. It is not stated if the manuscript used was the illustrated one, but it is tempting to think that this was indeed the case, the emperor taking with him the freshly completed portions of a work very close to his heart.

It must be emphasised, however, that it rests upon the assumption that Badāyūnī's statement that the manuscript took fifteen years to complete is correct; our notion would require revision if that statement proved to be inaccurate.

The date also appears to be supported by the style of the two dated manuscripts which were painted while the *Ḥamza-nāma* was in progress, namely the ʿĀs͟hiqa of 1568 and the *Anwār-i Suhailī* of 1570. These would have been done six and eight years after the *Ḥamza-nāma* was begun, which would be quite consistent with their mature workmanship. It also puts the completion of the *Ḥamza-nāma* five years before the commencement of the *Razm-nāma* which was painted between A. H. 990–93/1582–5, which is also what one would expect on grounds of style.[51]

The *Ḥamza-nāma* is a unique achievement displaying within itself all the essential qualities of the Akbar school in a clear manner. What is surprising is that it should do so at the very opening stages, so that we get the impression of a school fully formed at birth, as it were, carrying within itself all the inherent potentialities which it was left for later works to realize and elaborate upon. One is immediately put in mind of a similar occurrence in the history of Indian art, almost eighteen hundred years before the birth of the Mughal school when the sculpture of the Maurya dynasty seemingly came into existence all at once, and also seemed to be completely mature. Persian influence on both Maurya and Mughal art is undeniable, of the art of Persepolis on the one hand, and of the Safawi court on the other; but whatever was borrowed was subjected in each case to a similar process of rather quick and revolutionary transformation, the elegantly linear, static and decorative forms of the one being quickly transmuted into the plastically conceived and dynamic forms of the other. As one looks at the *Ḥamza-nāma,* the indebtedness to painting as practiced in the great atelier of S͟hāh Ṭahmāsp, from which at least two masters were directly recruited, is obvious enough to need no demonstration. At the same time, the differences between the two schools are of the most essential kind, the style of a work like the *Ḥamza-nāma* being inconceivable in Persia. Rather, the modelled figures moving in closely inter-knit and flowing rhythms find a conceptual parallel in ancient Indian wall painting done a thousand years before, as though the artists had

51. Gray, *Painting of India,* p. 83, states, without giving references, that the translation of the *Razm-nāma* was begun in 1582 and completed in 1589, and that the Jaipur manuscript was illustrated between 1584–9. This appears to be incorrect for Badāyūnī states that the work was begun in A. H. 990/ 1582, and in an early copy dated 1599 in the British Museum (OR 5642, Rieu, *Catalogue* I, pp. 57–58), Naqīb K͟hān, one of the principal translators states that the work took one and a half years, and gives the date of completion as S͟haʿbān 992/ 8 August–5 September 1584. This taken together with the death of Dasavanta, one of the principal painters of the Jaipur manuscript in 1584 (*Akbar Nāmā* III, p. 651), and an unpublished painting dated 18 Rabīʿ II 993/ 19 April 1585, a photograph of which is in the Victoria and Albert Museum, London, would indicate that the *Razm-nāma* was produced between 1582 and 1585. Skelton, "Harivamsa Manuscript," p. 48, refers to a miniature which "appears to be dated A. H. 994. If he is correct, the date of the *Razm-nāma* could be extended, perhaps, to 1586.

once again tapped the hidden springs of a rich and vital artistic imagination which had gone underground to lie dormant and forgotten for long centuries.

The connection, spanning a large period of time, is hardly susceptible of proof.[52] What is more clearly demonstrable, however, are traces of Indian pre-Akbar idioms, such as paintings of the *Caurapañcāśikā* group, which, though vestigial, are nevertheless present, and come to the surface sometimes in the vivid sense of colour, the rhythmical outlines of narrow-waisted, full-breasted women (Pl. 4; Glück, Taf. 41; *Ḥamza,* Pl. 48),[53] the heavy square-jawed heads (Pls. 16, 27), crowded movement, and in stray ornament like the mauve-and-maroon chequer pattern (Pl. 15; Glück, Abb. 6; *Ḥamza,* Pl. 4), the embroidery design of women's blouses (Pl. 2), and fleetingly, in the stylized foliage of a few trees (Glück, Taf. 48). Elements of the Indo-Persian styles and the Bombay *Candāyana* group are also present, but more difficult to detect, for several of these are shared with the Safawi style. Thus, the occasionally seen pale ground with sparse tufts of grass, the bricks shaded at one end, or the pure gold sky may be equally well derived from Indo-Persian, the Bombay *Candāyana* or Safawi traditions proper. It is, however, worth noting that a deep pink floral pattern, which I have not yet found in any other pre-Mughal painting but those of the Bombay *Candāyana* group (Pl. 108), occurs in at least one painting of the *Ḥamza-nāma* (Pl. 29, pattern on the umbrella; and Glück, Taf. 24, *Ḥamza,* Pl. 30, trappings of the horse to the right), and is symptomatic of the contact between the Bombay *Candāyana* group and the *Ḥamza-nāma.* But here, as in the case of the *Caurapañcāśikā* group, and perhaps to an even greater extent, the *Ḥamza-nāma* has quite absorbed and transformed what it has received in its outburst of creative activity, leaving behind but a few faint memories, which were not noticed previous to the discovery and careful study of the several types of pre-Akbar painting that have been coming to light in recent years.

Though the *Ḥamza-nāma* was executed over a period of fifteen years, it seems to show little stylistic development over this period of time. This may, perhaps, be more apparent than real, and due to the small proportion of paintings which have survived. But even if we compare paintings from the earliest volumes and those from later ones where we might most expect differences, we discover that these are on the

52. We have always wondered at this curious affinity, intuitively felt rather than directly demonstrable. Could it be that there were indeed surviving in the sixteenth century examples of ancient painting which are not available now but which were accessible to the emperor and his artists? For what it is worth, we discovered an intriguing reference to Akbar campaigning in the vicinity of Bagh, a site which preserves examples of fifth century wall painting to the present day, during the 1564 campaign in Malwa, *Akbar Nāmā* II, p. 346.

53. Gray, *Art of India and Pakistan,* p. 91, Colour Plate E, was the first to draw attention to a small detail showing a group of women at a well in a *Ḥamza-nāma* painting as being derived from what we call the *Caurapañcāśikā* group.

whole very slight.[54] Such variations as there are can be equally well understood in terms of the varying capabilities of artists or the exigencies of joint production, and appear to be no greater in range than what is to be observed within any individual volume of the *Ḥamza-nāma* or in any other large manuscript such as the Patna *Tārīkh* or the Victoria and Albert *Akbar-nāma* which were completed in a far shorter space of time.

A striking feature of the *Ḥamza-nāma* is its constant preference for rounded forms. Modelling in colour is constantly resorted to, and is reinforced by its interplay with the profuse, flat geometrical ornament that covers large areas with exquisite patterns. But however wonderful this ornament may be, it is not the artist's primary interest, his first concern being human life. The expression is not allusive, but direct and forthright, filled with harsh reality and psychological insight, evoking an urgent and anxious response. This is further intensified by the constantly restless motion, so vivid that it dissolves the very surface on which it is painted. Everything is swept up by this motion, whether animate or inanimate, as though impelled by a hidden inner energy. Rocks swell out of the earth, boulder tumbling over boulder, crag over crag; the waters swirl and twist in dizzying eddies; foliage bursts out of trees like light from fireworks. And all of this movement is not undirected and haphazard, but controlled and rhythmical, flowing swiftly but surely, submitting the surging turbulence to its own discipline, knitting all parts of the composition into an inseparable unity and providing them with structure and stability.

These are the essential characteristics of the *Ḥamza-nāma* style. If we turn to its individual components, we also notice typical and characteristic features. In the treatment of nature, we see the massive rocks, with swelling contours, shaded with washes of pink, mauve, blue and other colours, often splashed with white highlights and bordered by a heavy line (Pls. 12, 20). Shrubs and small trees, bereft of leaf or flower, are commonly placed in the interstices of rocks (Pl. 20). Trunks of trees are also heavily modelled, and painted in the same technique as the rocks, care being taken to render the texture of the bark, all skilfully enough handled so as to suggest the sap that flows within (Pls. 3, 21, 23, 31, and 34). The energy of this lifegiving substance is carried into the rendering of the thick growth of leaves, which dance and shimmer under its impact, forced from the centre to the outer edge of the foliage where they are reduced in size (Pl. 21). Each leaf is veined and rimmed with yellow, which is sometimes strong and bright and sometimes subdued, but almost

54. The painting depicting events at the birth of Muḥammad, which according to Glück belongs to the first volume of the *Ḥamza-nāma* (Glück, Abb. 1) thus differs but little from Glück, Taf. 7 which belongs to the eleventh volume.

always present. Flowers are seldom shown on these trees, though touches of colour are provided by the pink of fresh leaves (Pl. 21). There is present, however, a kind of shrub which is almost all flower, of a clearly Persian derivation, and flowering plants growing on the ground are abundant (Pls. 10, 21). A very distinct type of tree possesses a feathery foliage, the leaves being lightly daubed in green, yellow, blue and touches of red and brown (Pl. 23). These are sometimes organised in an ornamental pattern of intersecting circular forms, dotted blue and green at the base, and stippled with yellow at the edges (Pl. 20). This broad stippling technique, to a greater or lesser extent, is also used in depicting the lush banks of grass that cover the ground, a favourite device of the *Ḥamza-nāma* painters (Pl. 31). Rarely is the ground stippled with tiny green dots, and sometimes it is rather bare with sparse tufts of grass painted on a light pink, mauve or blue ground (Pl. 2) reminiscent of Persian and also work of the Indo-Persian and *Candāyana* groups (Pl. 99). The sky painted a uniform blue or gold (Glück, Taf. 42; *Ḥamza,* Pl. 50), or adorned with clouds in the Chinese fashion (Glück, Taf. 42), is also reminiscent of these traditions, but is rare, and is more often fluffed with streaks of white clouds to present a relatively naturalistic appearance (Pl. 26). The characteristic treatment of water in swirling eddies is derived from Safawi idioms (Welch, *King's Book of Kings,* Pl. on p. 86), but is now flecked with foam and, like all things in the *Ḥamza-nāma,* tumultuous in movement (Pl. 1). The older convention of representing water by a basket pattern is sometimes found, but its tight design is often in the process of breaking up in the direction of the new rendering (Pl. 10). The water is inhabited by large fish, leaping and diving, as well as turtles and crocodile-like *makaras* (Pl. 1). Animals on land are naturalistic and lively, painted with much sympathy (Pls. 23, 31), and the same is true of birds, notably the large herons nestling in trees (Glück, Abb. 21). The scene, however, is dominated by human beings filled with immense vigour and life. Bodies are modelled, both by colour and line, and the heads sometimes have the quality of portraits. The female figures, presenting an Indian ideal of beauty, are very distinctive, and are clearly derived, sometimes fairly obviously, from the *Caura-pañcāśikā* group, the similarities extending to small details like the ornamental pattern on the blouse and details of jewellery such as ornaments of the hair with flowers at the back of the head, ivory-peg earring, necklaces, armlets and bracelets (Pls. 2, 4, 15). These are, of course, not direct quotations from the *Caurapañcāśikā* group (Pl. 83), but naturally modified, though often the similarities are startlingly close.

The architecture, particularly the heavy ornamentation of geometrical tile-patterns as well as arabesques which extend to the domes (Pls. 5, 26, 32) is strongly reminiscent of Safawi painting, but is not as tightly and precisely rendered, the ornament sharing some of the vigour and movement of the style. The buildings are also often given considerable depth. A cityscape with a dominating gate is often seen in the background, and has a variety of Indian structures (Pls. 17, 33), which are often lavishly decorated, as are the gorgeous tents (Pl. 8). The clothes worn by men are frequently plain and free of decoration, but are also adorned with delicately drawn patterns. This strong emphasis on decoration of uniformly good qualitiy is probably due to the efforts of specialists in illumination *(naqqāsh)*, and does much to obscure the variety of *Ḥamza-nāma* painting. It is also peculiar to it, and is considerably reduced in miniatures of the subsequent phase.

Other manuscripts of the Ḥamza-nāma period

While work on the *Ḥamza-nāma* was going on, at least three other manuscripts were painted which further amplify our knowledge of Mughal painting in this period. The 1568 *ʿĀshiqa* (Pls. 35, 36) and the 1570 *Anwār-i Suhailī* (Pls. 39–42), both of the de luxe type, are two of them, bearing the same relationship to the *Ḥamza-nāma* as does the 1595 *Bahāristān* of Jāmī to, let us say, the Victoria and Albert *Akbar-nāma.* The third, which belongs to the class of singly painted manuscripts, is the *Astrology* manuscript in the Raza Library, Rampur (Pls. 43, 44). Miniatures from this manuscript have been mounted in a late album, so that there are no ascriptions now present, nor is there any date; but the style parallels that of the *Ḥamza-nāma* and easily places it in the same period.

The 1568 *ʿĀshiqa* and the 1570 *Anwār-i Suhailī,*[55] as mentioned above, are manuscripts of the de luxe type, executed in an exceedingly rich and refined manner, compared to the bold and forthright rendering of the *Ḥamza-nāma.* The *ʿĀshiqa* is particularly sumptuous with a fine *shamsa* on the title page, and a lavishly illuminated double page frontispiece. The margins of every page are additionally decorated with very carefully painted floral arabesques in gold. The two miniatures (Pls. 35 and 36) retain almost every feature of the *Ḥamza-nāma:* the elaborately decorated architecture, the tiled floors, the gardens in the background with luxurious banks of grass, streams with rippling water, leaves of trees fluttering with a life all their own, the plastic rendering of bodies, the brisk movement exemplified in the groom restraining the rearing horse, or the dancing fairies, these are all there; but the expression is more studiously delicate, the line more precious, the passionate outburst of the *Ḥamza-nāma* here finding a calmer and more controlled expression. A similar mood pervades the 1570 *Anwār-i Suhailī* (Pls. 39–42), where the colour is even more restrained, the leaves with bright yellow rims, for example, being no

55. See *supra* p. 59, fn. 25 for the contents of the colophons of these manuscripts.

longer in favour. The drawing is even more meticulous and careful, the sward and the feathery foliage of trees more deliberately stippled, and the lush banks of grass more minutely brushed, compared to the free and impressionistic handling in the *Ḥamza-nāma* (Pl. 41). The rocks are also not as massive and voluminous, lacking the heaving, surging weight of those in the *Ḥamza-nāma* (Pl. 41). It is as though the *Ḥamza-nāma* manner, in this deliberate and refined rendition, had lost some of its verve and fire, this aspect of the style anticipating future development. In this sense also, the 1570 *Anwār-i Suhailī* in particular, though earlier than some *Ḥamza-nāma* painting, pertains to the fully mature and later phases of the *Ḥamza-nāma* style as a whole.

Two miniatures in the 1570 *Anwār-i Suhailī* are done in a strikingly Persian manner, not in the Safawi style of Tabriz, but more in the manner of work found at Bukhara and in manuscripts of the Indo-Persian group. One of them, representing a king riding out on a hawking expedition (Pl. 42), is almost certainly the work of S̲h̲aḥm Muẕahhib, illustrations by whom are also found in a *Gulistān* of Saᶜdī in the British Museum (Pls. 37 and 38), a manuscript copied at Bukhara in 1567,[56] but, according to us, illustrated in India, probably about the same time as the 1570 *Anwār-i Suhailī*. This kind of artist was not able to adjust himself easily to the early Akbari manner and the type of work he represents died out at the court.

The Rampur *Astrology* (Pls. 43 and 44) shares many features with the 1570 *Anwār-i Suhailī*.[57] There is a similar perception of depth, greater than what is generally observable in the *Ḥamza-nāma*, receding distance being clearly suggested; and the modelling is more careful and detailed, particularly in the rendering of human figures. The compositions also are often rich and complex, and a similar sense of movement, brisk enough, but lacking the overwhelming vigour of the *Ḥamza-nāma*, is present. The foliage of trees, for example, is handled with the same clarity and restaint as in the 1570 *Anwār-i Suhailī*; the colour is as subdued, leaves with bright yellow rims, thus, being generally avoided (Pl. 44). The same careful draughtsmanship is visible in the painting of the banks of lush grass, or the tree with feathery foliage, and here again the colour is more lightly applied, the use of bright yellow being considerably subdued. Both in the Rampur manuscript and the 1570

56. Robinson, ***Descriptive Catalogue***, pp. 136, 127. The manuscript was copied at Bukhara by Mīr Ḥusain al-Ḥusainī in A. H. 974/1566–7. It has thirteen miniatures, seven in the late Akbar style and six in a style which Robinson believes to be of Bukhara, four of them signed by S̲h̲aḥm Muẕahhib. I suspect that S̲h̲aḥm though a follower of the Bukhara style, was probably a painter of the Indo-Persian group employed in the Mughal atelier, and he executed the six miniatures after the arrival of the unpainted manuscript in India, the seven others being done later on.

57. For reproductions of several miniatures and their description see Khandalavala and Mittal, "MS. of Tilasm and Zodiac," Figs. 1–35, pp. 17–20.

Anwār-i Suhailī, the archaisms seen surviving in the *Ḥamza-nāma* are more or less absent. But at the same time, we must remember that the Rampur manuscript is not of the de luxe type, but rather belongs to that class of manuscripts referred to earlier, also done by single painters, but more freely and rapidly executed. This accounts for the differences between it and the 1570 *Anwār-i Suhailī,* and the visual affinities with the *Ḥamza-nāma,* which are stronger at first sight than upon close examination. Its date, thus, should be around 1570, and it represents the mature style of the *Ḥamza-nāma* period at a freer and bolder level of expression than the 1568 *ʿĀshiqa* or the 1570 *Anwār-i Suhailī.* It is certainly not contemporaneous with the early phases of the *Ḥamza-nāma* as thought by some,[58] but is rather an achievement of the middle and late phases of the *Ḥamza-nāma* style.

Manuscripts of the post-Ḥamza-nāma period

De luxe manuscripts of the type represented by the 1568 *ʿĀshiqa* and the 1570 *Anwār-i Suhailī* are not found subsequent to the *Ḥamza-nāma* period until 1588, when we get a beautiful *Dīwān* of Anwarī with fifteen miniatures now in the Fogg Art Museum[59]. This is not to say that manuscripts were not painted, but they may well have perished, or remain to be discovered. The Fogg *Dīwān* was followed by a copy of the *Khamsa* of Niẓāmī dated 1595 which once possessed forty-four miniatures and is now divided between the British Museum and the Walters Art Gallery, Baltimore,[60] after which de luxe manuscripts are found frequently.

The tradition of manuscripts with singly painted miniatures as seen in the Rampur *Astrology* manuscript is continued in the post *Ḥamza-nāma* phase in works like the Beatty *Ṭūṭī-nāma* (Pls. 47–61) and the British Museum *Dārāb-nāma* (Pls. 45 and 46) which have been briefly discussed earlier.[61] Both of them are distinguished by their bold workmanship and considerable vigour, but the style is clearly more advanced than the Rampur manuscript, and memories of the *Ḥamza-nāma* style are sufficiently vestigial to allow us to date them around ca. 1580, just before work began on the *Razm-nāma* executed between 1582 and 1584, the next great undertaking after the *Ḥamza-nāma.*

In a previous chapter we considered the types of painting that existed in India before the advent of the Mughal style, and we have now briefly discussed this style in its opening and formative phases, as represented by the *Ḥamza-nāma* and related

58. *Ibid.*, p. 11.

59. See Welch, *Art of Mughal India,* pp. 26, 162, and Pls. 4a–d. The manuscript, according to the colophon, was painted at Lahore in Zuʿl-qaʿda 996/September–October 1588.

60. Welch, "Akbar's Khamsa," pp. 87–96 and Martin, p. 81, Plates 178–81.

61. See *supra,* p. 60.

manuscripts. Before the discovery of the *Ṭūṭī-nāma,* the link between this early Mughal painting and the indigenous schools flourishing before its advent was difficult to demonstrate, for not only was our knowledge of pre-Akbar painting fairly rudimentary, but even what we did know provided no natural link to the unique style developing at the Mughal court. As early as 1953, Basil Gray, controverting Khandalavala's position, had stated that the Mughal school "could not have happened without the pre-existence of one or more schools of painting in India,"[62] having already suggested the influence of the *Caurapañcāśikā* group in the formulation of the Mughal style by a comparison with a detail depicting women at a well in the *Ḥamza-nāma (Art of India and Pakistan,* Col. Pl. E). The evidence, however, was inconclusive, and the argument quickly degenerated into which came first, the chicken or the egg. We now know much more about pre-Akbar painting, though a great deal has yet to be understood, and, what is equally important, we also have a clear connecting link between the pre-Akbar and the Mughal style in the *Ṭūṭī-nāma* manuscript acquired by the Cleveland Museum of Art in 1962. This manuscript, to repeat, lays bare the actual process by which the Mughal style evolved out of a rich and varied background, and it is now, at last, that we turn with some confidence to its detailed consideration.[63]

62. Gray, "Development of Painting in India," p. 20.
63. A preliminary note on the manuscript was published soon after its acquisition. See Lee and P. Chandra, pp. 547–54. The present study has developed from work initiated at that time and can be read in conjunction with what follows.

CHAPTER IV
THE CLEVELAND ṬŪṬĪ-NĀMA

The illustrated manuscript of the *Ṭūṭī-nāma* of Ẓiyā' al-Dīn Nakhshabī in the Cleveland Museum of Art (No. 62.279, gift of Mrs. A. Dean Perry) once consisted of 341 folios. Of these, six folios (118, 141, 173, 266, 319, and 326) are missing, their present whereabouts being unknown. Seven folios (190, 205, 212, 235, 238, 247, and 286) are known to be in private collections.

The thin paper is of a light ivory colour, and the present size of the folios including the margins is approximately 20.5 by 14.5 cms. The written surface, which measures 16.5 by 10.5 cms., is enclosed by black, gold, and red rulings, and contains thirteen lines written in good *naskh.* The opening folios (1–7), a portion of fol. 340, and the last folio (341) are replacements of damaged or lost pages by a somewhat later hand, the writing as well as the paper showing an attempt to match the original, though the scribe writes fifteen lines to the page, except on the last page (341) where he reduces the lines to thirteen, apparently in an attempt to fill up the entire writing area. Seventeen other folios (156–7, 187–8, 251–3, 280–81, 285, 295, 297–8, 312–313, 331, and 334) are also replacements, but they are written on white paper in an indifferent *nastaʿlīq,* and appear to be of the nineteenth century.

The manuscript is complete except for the last fifteen couplets of a poetic epilogue that originally consisted of twenty-nine couplets. The first fourteen are written on fol. 341 recto, the verso side being left blank. As fol. 341 is a later replacement, it would appear that the last folio of the manuscript containing the missing fifteen couplets and the colophon, if any, was already lost when the replacement was made, or otherwise the text would have been continued on fol. 341 verso, which is instead treated as an end page carrying two smudged seal impressions, quite undecipherable, and a note at the bottom which records the despatch of four silk robes to one ʿAbd al-Wahāb Khān by somebody called ʿAbdullāh, and some scratchy drawings of floral plants by an idle hand. A fly-leaf at the beginning of the *Ṭūṭī-nāma,* also a later addition, carries an inscription written in *nastaʿlīq* giving the title of the book and the name of the author. It goes on to state that the manuscript was written in the *naskh* style, contained fifty-two tales with illustrations, and was a complete and authentic copy.

The miniatures are of varying sizes, but none are larger than the surface reserved for the writing. Most of the manuscript is in the Cleveland Museum of Art, and contains 211 miniatures while the seven folios known to be in private collections are all illustrated, making for a total of 218. We do not know if the six missing folios were also illustrated, but in view of the fact that folios with miniatures are known to have been sold separately, it is reasonable to assume that they were, making a total of at least 224 miniatures. It is also likely that the replaced folios possessed illustrations, but they are now lost, no attempts being made to provide substitutes when they were rewritten.

The present European binding in green leather was done by Rivière, and dates to the late nineteenth century. The manuscript was repaired and rebound at least twice previously, and all the folios have been recut and remargined, to a greater or lesser extent, with strips of paper. This undoubtedly accounts in large part for the destruction of several of the ascriptions to artists which were written on the bottom (fols. 32v, 37v, 58r) and also on the outer and inner margins (fols. 50v, 33v; and fols. 59v, 60v, 61v, 71v), the latter being a particularly vulnerable spot. It must be admitted though that several folios have sufficiently broad original margins, but do not for some reason possess ascriptions. The ascriptions to Dasavanta and Basāvana (fols. 32v and 33v) are in large and clear *nastaʿlīq,* and are preceded by the usual word *ʿamal,* "the work of." In ascriptions to other painters it is only their names that have been recorded (fols. 50v and 67r).[1] These are in quite cursive characters, close to, if not actually on, the framing lines, and obviously entered by another hand. We know of a few instances, for example the Jaipur *Razm-nāma,* in which precautions were taken to rewrite the ascriptions near the framing lines when a manuscript was rebound, and it seems that this practice may account for the peculiar character of these ascriptions. They may thus be later than the paintings, but their correctness, nevertheless, is vouchsafed by the style of the individual miniatures, and by the internal stylistic consistency of several miniatures ascribed to the same painter.

There are some instances when, though the ascriptions are present, it has not been possible for us to read them (fols. 68r and 70v), and several others where only a letter or so is visible or partially visible (fols. 37v, 42r, 43r, 46v, 62v, 69v, 323v), the rest having been trimmed away.

In addition to the twelve painters whose names have survived, it has been possible to ascertain the existence of several others on grounds of style. They number thirty-three, and have been named, for the sake of convenience, according to the letters of the alphabet (Painter A to Painter GG).

1. Both these forms are used in paintings of the Akbar period. Cf. for example, the British Museum *Dārāb-nāma* where fol. 19v has the ascription *ʿamal Bhagwān* while fol. 91v just has *Bhagwān.* Similarly the British Museum *Khamsa* of Niẓāmī dated 1595 has several miniatures in which just the name of the painter has been inscribed on the margin, see Martin, Plates 178, 180, 181.

Tables 2 and 3 give in summary form miniatures by the twelve painters arrived at on the basis of the ascriptions and style, and of the thirty-three other painters who have been identified on the basis of style alone. Painters contributing just a single painting have not been listed, and there are over sixty miniatures which we have not been able to relate closely enough either to each other or to previously ascertained work so as to warrant precise identification, though significant similarities have been noted in the descriptions.

A study of the miniatures reveals that the work of the various painters derives from all the three groups that we have outlined in the previous chapter on pre-Mughal painting, namely the *Caurapañcāśikā,* the *Candāyana,* and the Indo-Persian groups. But what is noteworthy is that the painters are unable to maintain the distinctiveness of their individual traditions for the most part, and we not only see them all striving in one way or the other to attain the Mughal idiom, but also freely borrowing elements from one another so that it is not always clear to which particular group an individual *Ṭūṭī-nāma* painter belongs. Thus, if the general composition and colouring is reminiscent of one group, we get a facial type, some landscape element, a patch of colour or an item of dress that belongs to another. The influx of Mughal ideals adds further complexities to the style and also contributes greatly to its variety and richness. In the treatment of nature, for example, such as the method of depicting trees, rocks, water, and sky, we see

Table 2

Ṭūṭī-nāma painters named in ascriptions

Banavārī 1	Fols.	49v, 50v*, 51v, 52v
Banavārī 2		87r*
Basāvana		33v*, 35r, 36v*, 207r
Dasavanta		32r, 32v*, 37v*
Ghulām ᶜAlī		66r*, 92r
Gujarātī		58r*, 59r*, 59v*, 61v*
Iqbāl		80v*
Lālū		71v*, 75v*, 78r, 81v*, 83v*, 84r, 84v*
Śravaṇa		67r*, 67v, 124r, 127r
Sūrajū		69v, 72r, 73r*, 79r*
Tārā 1		60r*, 95r, 102v
Tārā 2		63r*

* indicates folios with ascriptions. Those without are this author's attributions.

Table 3

Attributions to unnamed *Ṯūṯī-nāma* painters

Painter A	Fols.	20r, 30r, 46r, 46v, 54v, 55r, 55v, 56v
B		27r, 45r
C		29r, 158v
D		42r, 43r, 44v
E		62r, 62v
F		68r, 68v
G		70r, 70v
H		89r, 89v
I		97r, 97v
J		98v, 99v, 100v
K		105r, 125v
L		113r, 114r, 115v
M		128r, 128v, 129v
N		132r, 132v
O		143r, 160r, 161r, 161v, 168v, 170v
P		144v, 147r
Q		146r, 146v
R		154r, 179v
S		167v, 169v
T		177r, 177v
U		203v, 209v, 223v, 225r, 227r, 231v, 234r, 241r, 254v, 335v
V		213v, 216v
W		217r, 221r, 243r, 259v, 329v
X		222r, 222v, 229v, 267r, 275r
Y		239v, 248r, 262r, 268r, 268v, 271v
Z		219v, 235r, 238r, 240v, 247r, 291v
AA		258r, 261v, 263r, 269v, 321v, 323r, 323v
BB		282v, 287v
CC		292v, 315r
DD		293v, 308v, 325v
EE		306r, 307r
FF		316r, 324v, 332v, 333v
GG		328r, 339r, 340r

elements of pre-Mughal traditions existing simultaneously with features characteristic of the Mughal school, and this is also to be seen in the composition, colour, and the rendering of animals and human beings. In the depiction of the female figures, however, an overall preference is shown for a type derived from the *Caurapañcāśikā* group with rhythmical outline and characteristic patterns of dress and types of jewellery, but here also, some figures are more archaistic than others. This variety of stylistic features, the traditions to which they relate, and relationships to Mughal work have been emphasised in the descriptions of the various paintings, and we need not go into them further here.

The wide range of output is reminiscent to some extent of the diverse work found in early Mughal manuscripts, notably the *Ḥamza-nāma,* but more particularly in the British Museum *Dārāb-nāma* which is a manuscript of the same type as the Cleveland *Ṭūṭī-nāma,* having numerous miniatures painted by artists generally working singly, and in a bold and rapid manner. Like the *Dārāb-nāma* which has comparatively simple archaistic miniatures like the works of Ibrāhīm Lāhorī (Pl. 46) as well as the rich and sumptuous painting of Basāvana (reproduced in Welch, "Basawan," Fig. 1), the Cleveland *Ṭūṭī-nāma* has works ranging from the style, for example, of Painter A, to the kind of work represented once again by Basāvana or Painter U. Actually, the range here is even greater, and the artists can be divided into three broad categories, namely those still retaining to a large extent their native pre-Akbar manner, those who have made the transition to the Mughal style, and those who are in the process of making this transition to a greater or lesser degree. In the first category can be placed artists like Painters A, B, and Gujarātī, whose works are still very close to the pre-Mughal *Candāyana* group which was certainly their parent style, Mughal features being very few, though nevertheless present. Even in a miniature like fol. 59r by Gujarātī, almost a quotation from a painting of the *Candāyana* group, the more flowing rendering of the palms, the treatment of water, and the shaded green ground are Mughal features. Painter J is also to be placed in this archaistic category, but his parent style was that of the *Caurapañcāśikā* group.

Works done by the few painters in the above category can be effectively contrasted with the miniatures by artists who are working in the fully developed Mughal manner or a close approximation thereof. These are many more in number, and the workmanship of some of them is hardly to be distinguished from the style of the *Ḥamza-nāma* or paintings of that period. Painters H, L, R, S, U, W, X, Y, Z, AA, FF, Banavārī 2, Basāvana, Dasavanta, and Iqbāl are good examples. Not all of them possess the refinement of Basāvana, Dasavanta or Painters H and U, the work of Painters R, S, AA, FF, Banavārī 2, and Iqbāl being relatively coarser and heavier. Painters like V, GG, and Lālū, though competent enough, are still not at ease in the Mughal style; while the work of painters DD and EE is somewhat rough and ready. It is interesting to note that among the finest miniatures in this group are those ascribed to Basāvana and Dasavanta, the work being exactly what one would expect of these eminent artists at this early stage of their development.

A large number of painters attempting to make the transition to the Mughal style, but with varying degrees of success, have to be placed in a third category, for their work, though it has progressed beyond the archaisms of the first group, has not yet attained that degree of achievement in the Mughal style as is to be seen in the work of the second. Painters C, D, E, F, G, I, K, M, N, O, P, Q, T, BB, Banavārī 1, Ghulām ʿAlī, Śravaṇa, and Sūrajū can be assigned to this category. Their work is particularly fascinating for we see enacted in it the struggles that went on in the artist's mind as he tried to adjust his skills to the demands of a new patron and new ideals. We thus see fluctuations of the most startling kind as an artist relapses into what may well have been his native idiom in one miniature, while in another, he is back at work in a Mughal manner. Painter D, for example, is strongly reminiscent of the *Caurapañcāśikā* style in fol. 43, while in fol. 42r and 44v, his hand is much more Mughal, even though certain archaisms continue to be retained. And so also with Painter BB, who in fol. 282v presents a much more archaistic aspect than he does in fol. 287v. Painter O contributes a series of pictures ranging from fol. 143r, still very close to the *Candāyana* and Indo-Persian groups, to fol. 170v, in which the workmanship is almost fully Mughal. The mixture of styles at the Mughal atelier is also seen at its clearest in works of this group of artists, Painter G being particularly instructive. In the two paintings attributed to him (fols. 70r and 70v), we see a mixture of Indo-Persian, *Candāyana,* and *Caurapañcāśikā* elements, together, of course, with Mughal features.

For what it is worth, we can note here that conservative painters of our first category and intermediate painters working in a transitional style tend to play a dominant part in the first half of the book, while most of the painters working in the Mughal manner are found towards the end. It would be tempting to conclude from this that the manuscript shows within itself the entire gradual evolution of the Mughal style, from its early beginnings when it is still dependent on pre-Mughal elements, to its final phases when it has successfully achieved the fully developed style. But though this is symbolically true, it is not literally correct, for there are several exceptions, painters and paintings of the various categories appearing "out of sequence." Basāvana, Dasavanta, and Painter H, for example, occur in the earlier portion of the manuscript, and some works displaying archaistic characteristics occur in later portions of the book, notably those of Painter BB and fol. 278. Besides, there is reason to believe that the *Ṭūṭī-nāma* was painted in a relatively short time, about a year or so (*infra,* p. 167), and though the Mughal style came into being rapidly, this did not happen in a year. We may also mention here that miniatures by the same painter often tend to come close together and sometimes in sequence, though this is not always the case, and we can see no definite pattern emerging. One thing, however, is quite clear. If both the recto and verso sides of a folio bear a miniature, they are always by the same artist.

The technique by which the pictures are executed is that of the other schools of the time. The preliminary underdrawing was done with a brush, over which was applied a thin coat of zinc white. Large areas of colour, including gold, were then laid in and followed by the painting of smaller areas, the colour being applied in successive washes and repeatedly burnished. The dyes are the same as used in Mughal painting, and are largely of mineral extraction.

The Cleveland *Ṭūṭī-nāma* is the most profusely illustrated manuscript of this work that we know. An incomplete copy in the library of A. Chester Beatty, Dublin, consists of 143 folios with 113 illustrations in the Mughal style of the Akbar period, ca. 1580 (Pls. 47–61). Several separately mounted folios from this manuscript are in the collections of the National Museum, New Delhi; the Bharat Kala Bhavan, Banaras; and a number of private collections.[2] The Bharat Kala Bhavan also possesses a few folios of yet another manuscript of the *Ṭūṭī-nāma* having nine illustrations (Pls. 113 and 114). These are quite rough and coarse, and are of provincial workmanship with some Mughal influence. They can be dated to ca. 1580 on analogy with a *Laghu-Saṁgrahaṇī* dated 1583 and painted at Mātarapura in Gujarat,[3] but seem to belong to some other provenance, probably a centre in north India, as they show clear traces of the *Candāyana* style, particularly in the knobbed bed-posts (cf. Pls. 112, 114), the pink pattern of the pillow (cf. Pls. 110, 114), and the colours, which, though coarse, include the same pale tones as are observed in the *Candāyana* group.[4]

The style of the Cleveland *Ṭūṭī-nāma* is quite unique, no other manuscript of this type being presently known. I am aware of only two miniatures that belong to the phase of Mughal painting represented by it. One of these is the *Cow with a calf* in a private collection,[5] while the other, showing two men embracing, is in the Topkapi Serai Library, Istanbul.[6]

A critical description of each of the miniatures of the Cleveland *Ṭūṭī-nāma* is given below. When using the words ascription, ascribed to, we are referring to the entries made on the paintings. When we assign a painting to a particular painter on grounds of style, the words attribution, attributed to, are used.

2. Gray, *Art of India and Pakistan*, pp. 93–94, 142, Plate 120; and Binney, *Catalogue Mughal and Deccani Schools*, p. 33.
3. See Moti Chandra and Shah, "New Documents of Jaina Painting," pp. 395 ff., Figs. 16–19 and Colour Plates V–VI. The manuscript is dated V. S. 1640/ 1583 and painted at the village of Mātarapura near modern Kaira, Gujarat. The painter's name is given as Govinda.
4. I am grateful to Mr. Robert Skelton for kindly drawing my attention to this manuscript and to Rai Krishnadasa for giving me facilities to examine and photograph the miniatures.
5. Welch, *Meadow*, pp. 94–95. The miniature has been correctly attributed to Basāvana and bears close similarities to the painter's work in the *Ṭūṭī-nāma*. Cf. particularly fols. 35, 207.
6. I have not personally examined this miniature but know of it through a colour transparency kindly shown to me by Mr. Robert Skelton.

Critical descriptions of miniatures

First night

Fol. 8r
Khujasta kills the pet myna who advises her not to be unfaithful to Majnūn, her husband.

Khujasta stands in the courtyard, hand raised as though she has just dashed the bird on the ground. In the background is a boat with sails, indicating the absence of Majnūn who is voyaging over the seas in quest of fortune. Beyond, in the far distance is a hazy cityscape. To the right is a building with a kiosk on the terrace in which there is a lady.

The narrow gold sky with wisps of blue clouds, the pale tones of the architecture decorated with floral arabesques, the white crenallations of the parapet against a black ground and the delicate line are features suggestive of the *Candāyana* tradition (Pl. 108); but the strong impact of the *Caurapañcāśikā* group is evident in the red ground, the figure of the lady with its rhythmical contours, the pattern on her skirt, the tassels at the edge of her *oḍhanī* and at her waist, and the characteristic mauve and maroon chequered bed cover with lappets (cf. Pls. 82, 84), a feature also carried into the *Ḥamza-nāma* (Pl. 15; Glück, Abb. 6; *Ḥamza,* Pl. 4). Among early Mughal features is the building with open porch, tiled dado, doorway flanked by windows and panelled eaves (cf. Pl. 5), as is the depth imparted by the diagonals of the compound wall and the bed. The Mughal style, however, is particularly evident in the toning down of the strong angular rhythms which would be observed in the drawing of the lady if she were more purely in the *Caurapañcāśikā* manner. This modification of style, particularly in the rendering of figures, is a constant feature of the manuscript. The soft and fluffy grass occurs frequently in the *Ḥamza-nāma* (Pl. 31), as does the treatment of water, rendered by light white strokes on a dark ground (Glück, Abb. 36; *Ḥamza,* Pl. 52), though swirls and eddies are more common.

According to an argument presented by Ananda Krishna ("Reassessment," pp. 261 ff.), the distant landscape seen in this miniature occurs on Mughal painting only from ca. 1575 onwards, and is due to European influences, this proving the late post-*Ḥamza-nāma* date of this manuscript. Moreover, he continues, while Akbar period painting from this time on shows a direct knowledge of European originals as can be seen in the *Razm-nāma* (Hendley, Pl. 64), this miniature does not, being "derived at second hand." Now the *Razm-nāma* miniature under reference presents no such direct knowledge of European architecture, nor are any parallels cited. In any case, a very striking parallel with the *Ḥamza-nāma* is to be seen in Pl. 9, so that his argument is proved to be without any basis. He also stresses the treatment of the sails of the boat of this and another miniature from the same manuscript (fol. 212v), which, according to him, possess a "detailed modelling" unknown to the *Ḥamza-nāma.* Once again, he is far from correct, the soft graded modelling being utilised not only in representations of cloth (cf. Glück, Taf. 30) too many to enumerate, but also in the painting of the sail as well (Glück, Taf. 13; *Ḥamza,* Pl. 19), Ananda Krishna also sees in the occurence and style of the "herons" on top of the house "clear proof of a more developed style than that of the *Ḥamza* stage" *(ibid.,* p. 261). Aside from the question of the validity of his argument, this is really quite absurd, for the birds are entirely a product of his imagination. All that is present are curiously shaped spots caused by chipped paint.

Fol. 10v

The merchant hears of his wife's unfaithfulness (above); the unfaithful wife performs penance by plucking her hair (below).

The incidents are represented in two registers separated by a broad tiled band. The three striped domes of the band, the pink ground with curving horizon and the gold sky with delicately drawn arabesque-like clouds are drawn from the *Candāyana* group (Pls. 108, 111, 107). The female figures, however, both in the compact rhythm of line, and in ornamental detail, like the projecting end of the tasselled *oḍhanī,* the circular pattern on the skirt, and the striped *paṭakā* with vermiculate patterns at the edge are close to the *Caurapañcāśikā* group (Pl. 84).

While the upper picture is by an artist trained in the *Candāyana* tradition, the lower picture is by one who has learnt his lessons well and is already a proficient master of the Mughal style. Notice the face of the grief-stricken woman, who is shown crying with her mouth open, (cf. Pl. 30), an exceptionally sensitive and naturalistic portrayal unknown in India outside the Mughal atelier. The birds, though not yet free from a certain stiffness, are also painted in loving detail. Nevertheless, traces of *Candāyana* idioms still survive in the streaky decorations of the lady's *pāijāmā* (Pl. 108) and the projecting ends of the scarf, which are, however, painted with a softer and more yielding hand.

Second night

Fol. 16r

The sentinel in the employ of the S̲h̲āh of Tabaristan prepares to sacrifice his son to the ghost of the S̲h̲āh's soul.

In the foreground is the sentinel, about to cut the throat of his son before a lady who is the personification of the Shāh's soul. Behind them, hiding near a tree, is the Shāh himself, observing all.

The composition of the painting, with its high horizon, the sky painted in curving bands of gold and blue, the main scene laid in a garden-like setting with plants, trees, ornamental shrubs and cypresses is derived from miniatures of the Indo-Persian group (cf. Pl. 103). Pronounced similarity to the *Ḥamza-nāma* is evident in the soft treatment of the low, grassy mounds (Pl. 31) and in the rendering of the foliage of trees (Pl. 18). The figures are naturalistic and delicately drawn, and the woman too has shed many archaisms (cf. fol. 10v). Even so, traces of earlier origins are preserved in such features of the *Caurapañcāśikā* group as the circular pattern on the skirt and the narrow *paṭakā* projecting sharply at the base (Pl. 84).

Third night

Fol. 20

The goldsmith and the carpenter inform the king of a dream in which the golden images plan to desert the city for lack of worshippers.

Attributed to Painter A. His work is basically in the style of the *Candāyana* group with only some attempt to conform to Mughal tastes. Other paintings by him are fols. 30r, 46r, 46v, 54v, 55r, 55v, and 56v.

The miniature is divided into two parts separated from each other by a register showing domed balconies and windows. In this and in other respects it is similar to fol. 10v, both being derived from the *Candāyana* tradition, though the sharp stylistic dichotomy between the upper and lower panels is not to be observed here. The temple is a simple structure, the roof decorated by five ornamental plaques. The crowned images conform to those found in the *Candāyana* (Pl. 106). The figures in the foreground, however, in contrast to the rest of the picture, are clearly modelled and enlivened with bright colour, indicating a movement toward the *Ḥamza-nāma* style which is also to be observed in the upper panel of folio 10v.

Cf. fol. 58r, ascribed to Gujarātī, which also has strikingly close affiliations to the *Candāyana* group.

Fol. 23r

The goldsmith judged: the bear cubs trained by the carpenter behave as though they were his sons.

The carpenter, robbed by the goldsmith, decides to avenge himself. He invites the goldsmith, his wife, and two sons for dinner, the scene shown in a pavilion in the background. Hiding the boys, he produces two bear cubs instead, declaring them to be the goldsmith's sons, magically transformed. The matter being referred to the judge, the previously trained animals nuzzle up to the goldsmith as though he were their father, and he is pronounced by the judge to be so.

The miniature shows relationships to the *Candāyana* group in the yellow brick pattern striped with red lines (*New Documents,* Pl. 170). The golden sky marked with a curious three dots-and-a-dash motif appears to be a further misunderstanding of the dots at the horizon (representing a rocky encrustation) seen in the *Niʿmat-nāma* (Pl. 95; cf. fol. 20r for the same motif). The colour, however, is quite bright, as in early Akbar period painting. The mauve and maroon chequer pattern of the *Caurapañcāśikā* group is again present being frequently used by the painters of this manuscript, and retained even when other features have disappeared.

Cf. a miniature depicting the same subject in the Beatty *Ṭūṭī-nāma* (Pl. 47), a manuscript of the same date as the British Museum *Dārāb-nāma* (ca. 1580). The style is similar to that of a group of artists from Lahore, such as Ibrāhīm Lāhorī, whose work also occurs in the *Dārāb-nāma* (cf. Pl. 46). In contrast to the Cleveland *Ṭūṭī-nāma,* the Beatty miniature, in spite of its conservative character, has few traces either of pre-Mughal or *Ḥamza-nāma* idioms, so that the Cleveland miniature would be clearly and considerably earlier. The moment represented is that of the judgment, not the dinner that eventually led to it, continuous narrative being thus avoided. The colour is less vivid and the figures more expressive, particularly in the representation of the startled goldsmith and the clamouring cubs.

Fourth night

Fol. 26v

The mendicant's wife deceives him with a soldier.

In order to prevent his wife from being unfaithful, an Indian mendicant transforms himself into an elephant and carries her on his back, but to no avail, for no sooner does he set her down to forage for food than she finds a lover.

The miniature can be compared to fol. 16r which it greatly resembles in composition and colour. The rigidity of the curving band of golden sky, however, is obscured by tree-tops projecting beyond the horizon, and by wispy clouds floating in from the strip of blue sky above, resulting in an effect one step closer to the Mughal style proper. The elephant, as it tugs at a branch seeking food, is naturalistically treated; and the feathery foliage of the several trees anticipates similar representations in the *Ḥamza-nāma* (cf. Pl. 23).

Cf. Beatty *Ṭūṭī-nāma* (Pl. 48), a later miniature which is based on the Cleveland example, the posture of the elephant as it breaks off a branch for food being very similar. Archaisms, however, are abandoned there, the mendicant's wife and the soldier are clearly separated from the elephant who is placed in the background behind a rocky projection and clearly away from the lovers. In the Cleveland miniature, the separation is barely indicated by the intervention of a small tree. The feathery treatment of foliage is also absent from the Beatty example, only a vague memory being preserved in the small trees of the distant horizon.

Fol. 27r
The soldier receives a garland of roses from his wife which will remain fresh as long as she is faithful.

Attributed to Painter B. Cf. fol. 45r which is by the same artist.

The miniature is similar to fols. 30r and 56v attributed to Painter A in its attempts to graft Mughal technique on a *Candāyana* base, but is even more archaistic in expression.

Fol. 29r
The two cooks, who attempt to seduce the warrior's loyal wife are trapped by her in a cellar.

Attributed to Painter C. Cf. fol. 158v which is by the same artist.

In style the miniature is not very different from fols. 16r and 26v, and is by a painter originally trained in some kind of Indo-Persian style and at a similar level of achievement. This is evidenced by the light arabesque at the base and the curving horizon with gold sky which here, as in fol. 23r, is marked by the three dots-and-a-dash motif. The sharp transition from the green ground to the golden sky is partially obscured by the tufts of yellow grass. The stark strip of gold sky is also toned down by streaks of blue, red, and white clouds.

One of the would-be seducers slumps dejectedly on the floor of the cellar, while the other falls headlong to the same fate. The cot, whose deceptively weak mesh was the instrument of this downfall, is skilfully painted. What is taking place inside the cellar is revealed in section, a device used elsewhere in the *Ṭūṭī-nāma* (fols. 47v, 49v) and also in the *Ḥamza-nāma* (Pl. 18; Glück, Taf. 15; *Ḥamza,* Pl. 21), but of decreasing frequency later.

The bed on the porch of the building to the left does not possess the depth seen for example in fol. 8r, and is at odds with the bolsters on it. Note the popular chequer pattern of the *Caurapañcāśikā* group on the carpet (cf. Pl. 82).

Fol. 30r
The two erring cooks, dressed as maidservants, fall at the prince's feet and beg forgiveness.

Attributed to Painter A.

The miniature is very similar to fol. 20r, showing the same degree of Mughalisation of a *Candāyana* idiom.

Fifth night

Fol. 32r

The parrot mother cautions her young on the danger of playing with foxes.

Attributed to Dasavanta.

In the foreground is a large mango tree, sprouting fresh leaves and flowers, the conventional treatment modified by a closely observed rendering of the leaves and shades of darkness lurking between the branches. The mother bird, perched on the edge of the nest, speaks to the baby parrot hovering near it. Other little parrots flit amidst the branches and perch on the trunk, leaning forward to talk to the infant foxes. One of them has landed on the back, while yet another alights near the paw of an animal. The mother fox is seated to the left suckling her young. In the background is a palm with drooping fronds, its trunk obscured by the soft foliage of a tree with bent trunk placed in front of it. The ground consists of banks of light yellow grass.

The painting is related stylistically to the type of work seen in fol. 16r, and to an even greater extent in fol. 26v, but the quality is incomparably superior. The brush is observant, light and sensitive, barely seeming to touch the surface of the paper, but capturing easily and effortlessly the loose and scraggly texture of the nest, the shapes of mango leaves, the springy weight of the curving palm fronds and the shimmering light and shadow dancing in the feathery leaves. The play of the parrots with the foxes is innocent enough, but the lurking danger of this unnatural relationship, on which the wise and stately mother parrot expounds, is not overlooked, surfacing in the sudden and ominous turn of the head of the fox on whose back a baby parrot has playfully landed. The beasts, with ravenous eyes, open mouths and red tongues contrast with the vulnerability of the guileless birds. A sinister and foreboding note is struck in the hungry expression of the sharp-toothed mother fox, whose mind turns to thoughts of food as she herself suckles her young. All this subtle and keen observation indicates that we are in the presence of no ordinary painter but a great master.

The miniature can be attributed to Dasavanta on the basis of similarities to the miniature on the reverse, which is ascribed to him (see note under fol. 32v, *infra).* The style is similar to a detail in a *Ḥamza-nāma* painting also depicting foxes in a landscape (Glück, Taf. 33, lower right corner; Pl. 31), which could be by the same artist, but is more likely to be by Basāvana in view of the fuller and more ample brushwork.

Fol. 32v

The wounded monkey bites the hand of the prince, his chessmate, in the presence of guests.

A partially cut ascription in *nastaʿlīq* on the lower margin reads *ʿamal daswant,* the work of Dasavanta.

Within a walled enclosure is a pavilion in which the monkey, bleeding from the head where he has been struck by a chessman, bites the hand of the prince who is trying to fend off the animal. There is consternation on the faces of the guests as well as on those of the servants cooking food in the garden. Against the golden sky is painted the foliage of a slender tree, stippled softly and lightly in blue, green, and yellow on an indigo ground. This tree is stylistically very close to that of fol. 32r, as is the dark green field with yellowish clumps of grass, subtly shaded in order to suggest the uneven surface. The tall and slender figures are small in relation to the size of the picture, a characteristic that occurs in other miniatures in which Dasavanta played a part, notably in the Jaipur *Razm-nāma* (Hendley, Pl. 15, *Manu's ark;* Pl. 88, *Feast at Dwārikā;* and Pls. 47 and 48 which is an extraordinary double-page composition showing the army in battle array with literally hundreds of tiny figures). This is a feature shared with Khwāja ʿAbd al-Ṣamad, Dasavanta's teacher, if paintings of a later date by the Khwāja can serve as a guide to what his earlier style may have been (Pl. 65).

The brushwork throughout is extremely fine, as exquisite and gentle as in fol. 32r, and particularly apparent in the delicate and soft drawing of the faces. Movement is also handled with masterly skill, as, for example, in the sudden lunge of the monkey, the startled reaction of the prince, and also the guests who scramble up to be of help.

Dasavanta had a great reputation in his time, being described by Abū al-Faẓl as having no peers. The bulk of the work ascribed to him consists of thirty paintings in the Jaipur *Razm-nāma,* but as all of these are joint undertakings, his part being that of the designer *(ṭarrāḥ),* it is difficult to form a really clear idea of his work. Hendley reproduces twenty-eight miniatures (Hendley, Pls. 6, 9, 12, 15, 24, 32, 33, 39, 47, 48, 54, 61, 63, 67–74, 78, 83, 86, 88, 100, and 108). The other two have not been published. The Patna *Tārīkh-i Khāndān-i Tīmūriya* has only one picture (fol. 2v), also a joint work. Since both these manuscripts are of about the same date (ca. 1582–4) all that we have are examples of his work at one particular period, making a study of his style that much more difficult. In spite of this, it is obvious that we have in him a painter of the most lively imagination and skill, and a master in the formulation of convincing original compositions of the most unusual and difficult themes. Fol. 32r and 32v are particularly important as they give us an idea of his unaided work in its early formative stages.

The names of three painters who worked with Dasavanta in the Jaipur *Razm-nāma,* Banavārī (Hendley, Pl. 12), Tārā (Hendley, Pls. 71, 78, 83, and 86), and Śravaṇa (Hendley, Pls. 54, 69, and 70) also occur in the Cleveland *Ṭūṭī-nāma* (Banavārī, fols. 50r and 87r; Tārā, fols. 60r and 63v; and Śravaṇa, fol. 67r;)

Fol. 33v

The monkey slain, his blood to be used as medicine for the ailing prince he has bitten.

An ascription written on the right margin in the same hand as on fol. 32v reads, *ʿamal basāwan,* the work of Basāvana.

The ill man, fanned by an attendant waving a scarf, lies on a bed, his pulse being taken by the physician seated on the floor. A servant has seized the monkey by the hind legs, while another has severed its head.

Convincing representation based on direct observation of life make for a painting of great power, its style confirming the ascription to Basāvana. The various figures and the garments they wear are fully modelled, as are the large curtains, the red colour of their lining immediately centering our attention on the sick man, who is a wasted and weary person, without hope, hovering close to death, the light in the eyes about to go out. This sensitive, empathic portrayal is, among Mughal painters, distinctive of Basāvana, and reaches its most moving expression in his painting of the death of Ānandagiri done thirty years later in the miniature belonging to the Victoria and Albert Museum *Akbar-nāma* (Pl. 62). The movement is vigorous, the line soft and subdued, often tending to vanish, as one can notice in the rendering of the mauve and maroon chequer patterns of the bed-cover. Spatial depth is reinforced by the broad bed, the billowing yellow sheet in which the sick man is bundled, the white scarf over the doctor's shoulders, and the curtain wrapped around the projecting eaves, stylistic devices of which Basāvana was particularly fond. The pavilions on the terrace are flat, in contrast to later phases of Basāvana's style when the architecture is also treated in great depth. Attention may also be drawn to the solid and stable composition of the scene depicting the slaying of the monkey. The triangular arrangement of the figures is steady in itself, but is further reinforced by the posture of the kneeling figure with legs spaced well apart. Held firmly, the monkey is rendered helpless, and its body slumps, life departing as the head is severed from the body.

There are many affinities to the *Ḥamza-nāma* style, in colour, the types of figures, their vigorous movement, and the treatment of architecture. Particularly striking is the shading of the curtains, forming a v-pattern, very similar to a curtain in the 1570 *Anwār-i Suhailī* (Pl. 40), and also closely resembling the way in which a cloth sack is rendered in the *Ḥamza-nāma* (Pl. 25).

A rendering of the same subject in the Beatty *Ṭūṭī-nāma* (Pl. 49) is dull and insipid in comparison with Basāvana's inspired work. The composition is similar, but both in the general conception and in execution, it has the quality of an unfelt copy, none of the drama of the incident being captured. The priority of the Cleveland version is obvious.

Fol. 35r
The hunter throws away the baby parrots, who pretend to be dead, and captures the mother. Attributed to Basāvana.

Leaving his trapping equipment and shoes behind, the hunter has climbed up the mango tree, a pleased and satisfied expression on his face. Gently lifting the net, he slips a hand inside, presumably to catch and throw away the baby parrots who play dead. To the left are trees with feathery foliage beneath which hides the fox that has deceitfully led the hunter to the birds.

Basāvana's characteristic emphasis on the painterly rather than the linear is clearly seen in the leaves of the large mango tree which are painted with an extraordinary softness, even their light yellow outlines being generally omitted, and in the foliage of the two trees to the left. The tree on top is painted in subtly graded tones of yellow, purple, and pink against an ambient of light of the same colours. The tree below is similarly handled but employs the more usual lightly daubed blue-green and yellows. Soft purplish lights streak the dark green ground on the right, and are in harmony with the colour of the foliage. The subdued and glowing tones convincingly capture the quality of light in a dark forest. In painting the cage on the ground, the folded nets resting on tall sticks, and particularly the net spread over the parrots, it is the gauzy texture of the whole that has been emphasised, rather than the criss-cross pattern of intersecting threads. The figure of the hunter wearing a heavily modelled scarf around the shoulder is characteristic of Basāvana, and is repeated in fol. 36v which is ascribed to him.

The painting makes an interesting comparison with fol. 32r by Dasavanta which is similar in theme and composition. The styles of these two great artists correspond to some extent, but Basāvana's brush is softer and "wetter," lacking the lightness and restraint of Dasavanta's. This is exemplified in the rendering of the mango trees, Dasavanta's leaves being ever so slightly crisper, and is also to be seen in the treatment of the trees with daubed foliage. Basāvana uses a more extensive variety of brush strokes, the colour is a little deeper, and the spots of paint merge and coalesce to a greater extent, as compared to the lighter, more orderly and neat brush of

Dasavanta. It is for these reasons that I believe the detail of a *Ḥamza-nāma* painting (Pl. 31; Glück, Taf. 33), showing foxes in a landscape, to be in all probability the work of Basāvana, rather than of Dasavanta. Attention is especially drawn to the banks of grass in the *Ḥamza-nāma* painting, the handling almost identical with Basāvana's tree with feathery foliage in the Cleveland miniature, and also to the trunks of trees in close and life-like embrace. The man scrambling up the rope in the *Ḥamza-nāma* is also strongly reminiscent of Basāvana's work both in the modelling of the figure and the closely observed rendering of the rope which adjusts its fall to the weight and pull of the man who is on it.

For other works by Basāvana in this manuscript, see fols. 33v, 36v, and 207r.

Fol. 36v

The hunter offers the mother parrot to the king of Kāmarūpa.

The partially cut ascription on the left margin reads *ʿamal basāwan,* the work of Basāvana. That two paintings identical in style (fols. 30v and 36v) are ascribed to the same painter confirms the authenticity of the ascriptions.

The king, dressed in white, sits on a stool placed on a carpet of rich blue within a building supported by columns with elaborately decorated shafts and capitals. The parrot is perched on his hand. The hunter, who faces him, leans forward to extol the virtues of the bird, holding the empty cage and the string secured to the bird's leg in one hand, and fondly stroking the birds tail with the other. Outside the room, the wall of which is painted in gold, is a seated row of three courtiers arranged diagonally, with a tiled pink wall behind. An unseen wind blows a swirling red curtain through a cusped arch placed in a projecting section of the wall, and wafts the branches of a tree to the right. The foliage of all the four small trees in the background is painted by softly daubed brush strokes, against a sky of deep blue.

The ascription of the picture to Basāvana is supported by the style of the miniature and its close similarities to fol. 33v (also ascribed to Basāvana) and to fol. 35r (attributed to Basāvana). The sky and stippled trees are identical in style, as is the figure of the courtier seated in the middle of the row, and the man holding the monkey's hind legs (Fol. 33v). The hunter wearing the thick and heavily modelled scarf, criss-crossed with stitches, is identical with the same person in fol. 35. The extremely rich and mellow colouring, the soft glowing brushwork, the constant attempt to give roundness and depth to objects by modelling in colour, all characteristic of Basāvana's work, are present, as is the swelling, expansive rendering of human and animal bodies.

This miniature is also interesting in as much as it reveals some traces of Basāvana's pre-Mughal manner which he had not yet fully outgrown, something to be expected at this very early stage in his career. One thus notices the archaic face of the man seated at the lowermost end of the carpet, and the somewhat awkward and cramped placement of the fly-whisk bearer behind the king, though the modelling and the details are delineated with the painter's usual skill. The white cornice with embattled parapet painted against a black ground, the tasselled valance with floral pattern in dark pink, the light pink wall with hexagonal tiles, the golden dots, crosses and arabesques decorating the *jāmās* have strong affinities to similar devices in the *Candāyana* group (Pls. 110 and 107), and perhaps it was this style that Basāvana practiced before he entered the Mughal atelier.

A painting like this can be contrasted with Basāvana's work of a decade later, as represented by a miniature in the British Museum *Dārāb-nāma* of ca. 1580 (Welch, "Basawan," Fig. 1). Though the same predilection for plastic form is to be seen in both, it is even more developed in the *Dārāb-nāma*. Not only have archaisms entirely disappeared, but there is experimentation with foreshortening. The architectural setting is also painted in much greater depth so that it is difficult to see why Basil Gray sees no difference in the date of the *Dārāb-nāma* and the Cleveland miniature (*Artibus Asiae* 28, p. 100).

The name Basāvana is quite popular in eastern Uttar Pradesh, though it may be known elsewhere, and if Basāvana did come from this part of India, it may suggest that paintings of the *Candāyana* group were possibly painted in that region. See, however, fol. 58r, *infra*.

Sixth night

Fol. 37v

The parrot addresses Khujasta at the beginning of the sixth night.

Attributed to Dasavanta.

A cut-off ascription is visible on the lower margin. It seems to read [*ʿamal da*]*s* [*wa*]*nt*, the work of Dasavanta.

The delicately brushed grass and the arabesque of the golden carpet are identical to those in fol. 32v, as is the sensitive drawing of the lady. The composition is simple, but the gently tilted cage and the eloquent gestures of Khujasta are the result of accurate observation, not found in the other miniatures of this subject in the manuscript. For a similar female type in the *Ḥamza-nāma*, see Pl. 4.

Fol. 42r

Seven men disputing possession of a woman bring her before the Tree of Justice into which she is absorbed.

Attributed to Painter D. Cf. fol. 43r and fol. 44v which are by the same artist. His work assimilates features both from the *Caurapañcāśikā* and the *Candāyana* group, and is evolving towards the Mughal style, though in this miniature the *Caurapañcāśikā* elements are fairly subdued. See remarks under fol. 43r.

The lady is barely visible, having almost completely merged into the flowering tree which is of pre-Mughal derivation, occurring also in the Mandu *Kālakācāryakathā,* the *Niᶜmat-nāma,* the 1516 *Mahābhārata* and other paintings (Pls. 67, 96, 73; also see Chapter V, p. 165). The disputants who consist of a carpenter, a goldsmith, a tailor, a hermit, a traveller, a prefect, and a judge are shown in two superimposed rows. The postures are stiff, the faces with rounded chins frequently repeated, little attempt being made to characterise them in any special manner except by providing a few with the tools of their trades. We thus see a bag around the shoulder of the traveller (top right) and a hammer and an axe tucked into the girdles of the two figures on the bottom left. The mango can be contrasted with representations of the same tree by Dasavanta and Basāvana (fols. 32r and 35r). The lavender ground, edged with low rocks, shaded pink, are of the *Ḥamza-nāma* type (cf. Pl. 2 and Glück, Abb. 25). The sky is painted gold.

The same incident is represented in the Beatty *Ṭūṭī-nāma* (Pl. 50). The ornamental tree has been replaced by a more realistic one, the high horizon eliminated, the figures painted more realistically and grouped together, and the rocky prominence is of a projecting, crystalline type popular in the British Museum *Dārāb-nāma* and later.

Seventh night

Fol. 43r

The parrot addresses Khujasta at the beginning of the seventh night.

Attributed to Painter D. Cf. fols. 42r and 44v which are by the same artist.

The lady stands against a green field with clumps of sprouting grass. The rocks on top are shaded yellow, instead of the more usual pink. On either side are pavilions containing beds with pillows. The wall of the pavilion to the left is pink, not unlike similar patches of colour found in works of the *Candāyana* group, but the one to the right is a bright red, immediately putting us in mind of the *Caurapañcāśikā* style. Further likenesses to it are found in the stiff ends of the *oḍhanī* with projecting tassels, the flat perspective of the beds, tassels suspended from folded curtains, knotted wall brackets, domed kiosks, and utensils on the floor (Pl. 84). The jewellery, including the ivory-peg earring, and the pattern on the blouse are also derived from paintings of the same style (Pl. 83). Even small details like the peculiar green and red spots interspersed with foliate v-shaped forms are almost identical to similar decoration in several paintings of the *Caurapañcāśikā* group (Pl. 84; also cf. fol. 70r). In spite of these archaic features, an attempt has been made to model the torso and the face, and to shade the sky with white clouds so that the hard and sharp-edged forms of the *Caurapañcāśikā* group are considerably softened.

The facial type is identical to that occurring in fols. 42r and 44v, and all three paintings are done by the same artist. This miniature demonstrates the intermingling of the various pre-Akbar idioms in the Mughal atelier, and the striking presence of the *Caurapañcāśikā* strain, often obscured later, except for the female figures and some ornamental patterns. The same figural type, right down to the peg-shaped earring and the distinctive pattern on the blouse, but further Mughalised, occurs in the *Ḥamza-nāma* (Pl. 2).

Cf. fol. 282v for another painting with strong *Caurapañcāśikā* memories.

Fol. 44v

The vizier dissuades the king of Bahilistan from executing the *darwīsh* who asks for his daughter's hand in marriage.

Attributed to Painter D.

The painting is very similar to fol. 42r, and particularly to fol. 43r by the same artist, the figural types, the faces and the composition being identical. The *Caurapañcāśikā* sources, however, are hardly noticeable.

Fol. 45r

The *darwīsh* brings in as dowry an elephant laden with gold.

Attributed to Painter B. Cf. fol. 27r by the same artist.

Fol. 46r

The *darwīsh* brings the King of Kings before the king of Bahilistan.

Attributed to Painter A.

An ascription on the lower margin has been cut off.

The painting is based on the *Candāyana* group as is indicated by the treatment of grass in the foreground (cf. Pl. 111), the lavender wall, the white ramparts on black ground, and the high horizon with stylised clouds. The colour, however, is a little deeper, and the pavilions on top are given depth. They were, however, once done in the *Candāyana* manner as in fol. 20r (central register), this being revealed by the chipped paint, and were changed to accord with Mughal taste. The faces, too, were refined, and the painting is an interesting illustration of the manner in which recruits were taught to adapt to changed preferences.

In spite of the different tones of colour and the more delicate line of the faces, the miniature seems to be by the same artist as fols. 20r, 30r, and 46v.

Fol. 46v

The king of Bahilistan offers his daughter to the King of Kings.

Attributed to Painter A.

Faint traces of an ascription, trimmed away in the course of rebinding, can be seen on the lower margin.

The miniature is very similar to fol. 46r, and shows fewer signs of retouching by another artist. As such, its proximity to fols. 20r and 30r is more evident, though the domed pavilions are in the Mughal manner.

The division of the main panel, which has two arched panels at the sides, is reminiscent of the *Candāyana* group (Pl. 107). *Caurapañcāśikā* features can be seen in the canopy ornamented with vermiculate pattern and strung in three loops with tassels in between (Pl. 84). The figures of the king of Bahilistan and his daughter, who bows to the ground, though expressive, are adapted from indigenous tradition (Pl. 81). The faces are heavy, but not unknown to the *Ḥamza-nāma* (Pl. 27), the door with its mother-of-pearl inlay to be also found in that manuscript (Pl. 25).

Fol. 47b

The Brahman gambler sees the daughter of the king of the jinns in a pit together with an old man and a cauldron of boiling oil.

The age of the old man is symbolically indicated by a beard, otherwise there is little to distinguish him from a youth. The front section of the pit has not been represented so that we may observe what is happening inside.

The painting is similar to, but apparently not by the same hand as fol. 49v attributed to Banavārī 1.

Fol. 49v

The daughter of the king of the jinns bows before the King of Kings who has just undergone the ordeal of passing through boiling oil to emerge as a youth.

Attributed to Banavārī 1 (cf. fols. 50v, 51v, and 52v).

The composition is similar to fol. 47v, but is more geometrically balanced, the drawing surer and steadier so that it is apparently the work of another artist. The feathery leaves of the tree are clear in the center, but illusionistically smeared towards the edges. The formal manner in which this has been done can be compared with the much freer and more convincing work of Dasavanta and Basâvana (fols. 32r and 35r). The faces particularly of the Brahman, whose vocation is indicated by a book, and that of the King of Kings are very close to some in the *Ḥamza-nāma* (Pl. 16), as is the sky with streaks of white clouds (Pl. 26).

Fol. 50v

The rejuvenated old man and the daughter of the king of the jinns take leave of the King of Kings.

Attributed to Banavārī 1.

An ascription on the left margin reads *banwārī,* (by) Banavārī.

The miniature is quite similar to fol. 49v, especially the figural types, and both are probably by the same artist.

Ascriptions written on paintings of the Akbar period give us three names, Banavārī, Banavārī Kalān (the elder), and Banavārī Khurd (the younger); but these probably represent only two painters, the appellations *kalān* and *khurd* being given when it was felt necessary to distinguish one from the other. Banavārī Kalān's unassisted works are found in the British Museum *Dārāb-nāma* (fol. 36r), and he is probably the same as the Banavārī who painted the *Viṣṇu Trivikrama* in the Jaipur *Razm-nāma* (Hendley, Pl. 147). He also assisted as the colourist in five other paintings in the same manuscript (Hendley, Pls. 12, 26, 27, 51, and 104). One painting done by him working individually is in the Patna *Tārīkh* (fol. 148r), while on another five miniatures in the same manuscript he worked as colourist (fols. 22v, 23v, 51r, 69r, and 69v). He continues to work in the *Akbar-nāma* as the colourist for six miniatures (Victoria and Albert *Akbar-nāma*, Nos. 11, 33, 48, 90, 98, and 104). A painting by him is also found in the Beatty *ʿIyār-i Dāniṣh* (fol. 83r), and we have come across no later works ascribed to him. His style tends to be archaistic and is to be easily distinguished from Banavārī Khurd's, who is more in the mainstream. The Banavārī, of the *Ṭūṭī-nāma* (fols. 49v, 50v, 51v, 52v), here called Banavārī 1, is most probably the Banavārī Kalān of later paintings. The Banavārī of fol. 87r may perhaps be Banavārī Khurd, here called Banavārī 2.

Eighth night

Fol. 51v

The parrot addresses Khujasta at the beginning of the eighth night.

Attributed to Banavārī 1.

The tufted mounds of grass are laid in a formal pattern, and at its edge is placed a mauve hill, shaded with white just at the top, the technique being similar to the golden hill of fol. 50v. The female figures of the two folios and also of fol. 49v are very similar, all three pictures being by the same artist.

Fol. 52v

The astrologer predicts a calamity for the newly born prince in his thirteenth year, but one which he would be able to overcome.

Attributed to Banavārī 1.

The mother, holding the newly born babe, sits behind the astrologer who is consulting his charts and papers. A singer and two musicians are arranged in a curved row.

The figures of the ladies and the men are similar to those in fols. 49v and 50v, and the miniature is by the same artist. Here, as elsewhere in the manuscript, female figures are patterned after *Caurapañcāśikā* types (cf. Pl. 82), even if the drawing is refined. Notice also the bright patch of red on the wall here, and on the bed in fol. 51v.

Fol. 54v

The young prince is presented to the king, his father, by his teacher, but refuses to speak.

Attributed to Painter A.

This and several succeeding pictures (fols. 54v–73r), which illustrate extensively the story of the prince who would not speak, are by very similar artists.

From the *Candāyana* tradition are derived the pale tones of colour and the characteristic pink pattern of the seat cover (cf. Pl. 108), the latter surviving in the *Ḥamza-nāma* as well (Pl. 29). The composition and the floral arabesques on the gold backdrop of the throne are features shared with the Indo-Persian group (Pls. 103 and 101).

Fol. 55r

The king's handmaiden takes the prince away to the harem.

Attributed to Painter A.

The figure of the king, with heavy cheeks and prominent chin, is very close to that of the King of Kings in fol. 46v, and the two paintings seem to be by the same artist. The composition, with its curving registers, date palm and cypress, and the finely drawn arabesque ornamentation, recalls the Indo-Persian group (cf. Pls. 104 and 102), but the pale colour and the cursive dot-and-check decoration of the background are reminiscent of the *Candāyana* group (Pl. 108).

Fol. 55v

The prince rejects the amorous advances of the king's handmaiden.

Attributed to Painter A.

The work is identical with that of fol. 55r. *Candāyana* traditions are evident in the decor of the bed with its blue and gold cover, the pillow with pink stripes (*New Documents,* Pl. 168), the floral motif on the canopy, the columns with light arabesque decoration, the pink wall with little flowering plants, the yellow arabesque of the entablature, and the yellow floor with lined rectangular tiles (Pl. 110 and 109). It is only the figures which are modelled and to that extent endowed with Mughal characteristics.

Fol. 56v
The prince being taken away for execution on the false complaint of the handmaiden.

Attributed to Painter A.

By the same artist as fol. 55v. Cf. also fol. 30r.

Fol. 57v
Landscape with a lotus pool.

The miniature immediately recalls in colour and mood a painting from the Bombay *Candāyana* (*New Documents,* Col. Pl. 24). The lotus pool, also found there (*ibid.,* Pl. 166), has a long tradition in pre-Mughal painting, occurring both in the Western Indian school and the *Caurapañcāśikā* group (*ibid.,* Pl. 71; and Pls. 74, 75, and 84). With the Western Indian style is also shared the motif of date palms with squirrels scrambling up their thin trunks, a feature that survives in the Rampur *Astrology* manuscript. Cf. fol. 59r for a similar miniature.

Fol. 58r
A woman asks her lover to leave her house, brandishing his sword and feigning rage in order to deceive her husband who has just arrived.

On the lower margin is a partially visible ascription which reads *kār guj* [*r*] *āt* [*ī*], the work of Gujarātī. For other paintings by the same artist see fols. 59r, 59v, and 61v.

The wording of the ascription is unusual, for Gujarātī is not a proper name, but designates a person from Gujarat. As such, it is seen as part of the name in ascriptions mentioning no less than seven Mughal painters: Bhīmajū Gujarātī (Patna *Tārīkh,* fols. 16v and 166v; and Beatty *ʿIyār-i Dānish,* Nos. 72 and 73); Devajū Gujarātī (Patna *Tārīkh,* fol. 143v); Premajū Gujarātī (Patna *Tārīkh,* fol. 62a; and Victoria and Albert *Akbar-nāma,* No. 102); Sūrajū Gujarātī (Patna *Tārīkh,* fols. 89b, 121a, and 61b); and Kesū, Śankara and Śivarāja Gujarātī (Beatty *ʿIyār-i Dānish,* Nos. 62 and 82; 78; and 30). None of their work seems to be related to the present miniature. In all of the above ascriptions Gujarātī follows the name, while here, it is clearly the first word.

The miniature is strikingly reminiscent of the Bombay *Candāyana,* paralleling it in many respects including the pale colour, the arabesque-like clouds, kiosks with striped domes and the various ornamental patterns (Pls. 107 and 109). The female type, as is so often the case, leans heavily on *Caurapañcāśikā* prototypes, and the figures are modelled in the Mughal manner.

The miniature is closely comparable to fol. 20r, and also to fol. 10v, though the lower portion of the latter picture is pronouncedly Mughal in character.

We have here a painter evidently from Gujarat working in the *Candāyana* manner, an area that is not usually associated with paintings of that group. This could either indicate the prevalence of the style in Gujarat, or that the painter had migrated and settled in the north.

Fol. 59r

Landscape with a lotus pool.

Ascribed to Gujarātī.

Ascription on the lower margin reads *aiẓan,* "the same," this being a reference to Gujarāti who painted the previous picture (fol. 58r).

The miniature, like fol. 57v, is close to *Candāyana* traditions, but the workmanship is smoother and more even-flowing.

See also remarks under fol. 57v, *supra.*

Fol. 59v

The prince, once reprieved, is returned to the place of execution a second time on the plea of the king's handmaiden.

Ascribed to Gujarātī.

The ascription on the left margin reads *aiẓan,* "the same," this being a reference to Gujarātī who painted the previous picture (fol. 59r). Both paintings rely heavily on *Candāyana* traditions, particularly fol. 59r, the small landscape of which is almost a direct quotation from the *Candāyana* group.

This miniature is very similar in composition to fol. 69v, which seems, however, to be by another artist.

Fol. 60v

The deceitful wife assaults her erring husband.

Attributed to Tārā 1.

An ascription on the lower left margin reads *tārā,* (by) Tārā.

Tārā is one of thirteen painters specifically named by Abū al-Faẓl as having attained fame. There is one miniature by him in the British Museum *Dārāb-nāma* (fol. 95v) and two more in the Jaipur *Razm-nāma* (Hendley, Nos. 37, 56), in all of which he works without assistance. He collaborates as colourist to Dasavanta in four other paintings in the *Razm-nāma;* and in one other painting of the same manuscript he is the designer (Hendley, No. 118). An artist named Tārā Kalān works as colourist to Basāvana in two splendid paintings of the Victoria and Albert *Akbar-nāma* (Nos. 17 and 61, Welch, "Basawan," Figs. 11, 15), and he may or may not be the same as this Tārā. It is surprising that Tārā was not the designer of more pictures in view of Abū al-Faẓl's praise, but this may be due to an insufficient number of surviving paintings. He certainly reveals himself to be a painter of considerable promise in the present example.

Though the painting shows vestiges of the *Candāyana* group, particularly in the muted colour and the faces of the seated men (Pl. 109), it is the Mughal features that are more apparent, particularly in the rendering of dramatic action. The man, taken unawares, falls back before the impetuously rushing lady who has gripped him by the collar while she tugs his beard. The gestures of the seated companions are very lively, the movement as a whole reminding us of the broad and expansive action to be seen in the *Ḥamza-nāma.* Also cf. fols. 95r, 102v, by the same artist.

Cf. Beatty *Ṭūṭī-nāma* (Pl. 51). The incident occurs in a tent instead of a building and the male attendants have been eliminated as inessential, but the composition is clearly based on the Cleveland miniature. The action is fairly restrained, even though the lady plucks the unfortunate man's beard with both her hands.

Fol. 61v

The handmaiden again appeals for justice and the prince is led to the place of execution for the third time.

On the left margin is a partially visible inscription which reads *gujrā*[*tī*], (by) Gujarātī.

The painting is very similar to fol. 58r, and is equally dependent on the *Candāyana* group. The figures painted against the pink hexagonal tiles (cf. Pl. 107), except for their dress, are almost untouched by Mughal mannerisms. The background of the lower panel, however, is filled with a grassy field as in Mughal painting.

The floral pattern done in deep pink on the valance above the handmaiden, a feature which occurs frequently in this manuscript, is a characteristic motif of the Bombay *Candāyana* (cf. Pl. 110), and we have not noticed it in any other type of pre-Akbar painting. It occurs once in the *Ḥamza-nāma* (Pl. 29) where its survival is significant, in as much as it hints at the not so apparent contribution of the *Candāyana* group to that manuscript.

Fol. 62r

The lover's son makes an elephant of the pastry dough carried by the unfaithful wife and puts it in her basket.

Attributed to Painter E. Cf. fol. 62v which is by the same artist.

The miniature is again based on the *Candāyana* group, as is seen in the yellow tiled floor, the lavender ground, the curving horizon, the blue and gold sky, the narrow-trunked leaning tree with stylised spray-like foliage (cf. Pl. 111), and the heavy-headed figures.

The Beatty *Ṭūṭī-nāma* has a miniature (Pl. 52) depicting the same event, and also including the one which succeeds it. The workmanship is more archaistic and shows greater proximity to the *Ḥamza-nāma* style than other miniatures in the manuscript, particularly in the figure of the woman and the mounds of yellow grass, and rather heavy shading of the rocks. In spite of this, the Cleveland miniature is considerably earlier than the Beatty version, exemplifying as it does the first hesitant beginnings of the Mughal style, while the Beatty miniature represents a well-settled idiom.

Fol. 62v

The unfaithful wife explaining away the presence of the dough elephant.

Attributed to Painter E.

An ascription in the lower left margin has been cut off.

The miniature is by the same artist as fol. 62r, though the artist has here adjusted to Mughal requirements to a greater degree.

The painting is of particular interest as it reveals the manner in which an artist originally trained in *Candāyana* idioms (cf. Pl. 111) transforms his style to conform with what is demanded of him in the Mughal atelier. The composition is still basically the same with a paved floor of brick-shaped tiles, a high curving horizon, a leaning tree with oval foliage, and blue sky with arabesque clouds. However, each of these features has been modified, the brick floor being finished in deeper colour, the figures and their clothes more clearly modelled, and the background more naturalistically treated to resemble a field with shaded mounds sprouting grass. The trunk of the tree is also given a more naturalistic shape, and the leaves are fluffed with blue, green, black, and yellow daubs. The *Candāyana* sky, done here in bands of light and dark blue, is also undergoing changes. The light blue strip is now thought of as distant ground, this being clearly indicated by the clumps of grass lightly painted on it in a slightly darker shade of blue; and the ornamental shapes of the clouds in the deep blue sky are considerably subdued and obscured.

Cf. fols. 146r, 150r, and 152r for similar compositions.

Fol. 63v

The handmaiden appeals for justice and the prince is taken to the execution site for the fourth time.

Attributed to Tārā 2.

An ascription, unusually placed on the miniature itself, just below the foliage of the tree, reads *tārā,* (by) Tārā. The style differs from another painting ascribed to Tārā (fol. 60v), so that either the ascription is incorrect, or it refers to another painter with the same name. Both a Tārā and a Tārā Kalān are known in the Akbar period (see note under fol. 60v).

The miniature is close to fol. 46v which is the work of Painter A. *Candāyana* features are apparent in the two registers with flat patches of colour, lightly drawn arabesques, and the high curving horizon against a golden sky (Pls. 106 and 111). The usual Mughal features in the treatment of the figures are present, but we also see strong *Caurapañcāśikā* survivals, especially in the stiff, angular outlines of the dress of the handmaiden and the yellow skirt decorated with circles containing foliate stars (*New Documents,* Col. Pl. 21, woman lying in pavilion). This ornament is also found in other miniatures of this manuscript (cf. fol. 70v) and continued in the *Ḥamza-nāma* (Pl. 19), as is the bright ornamental foliage of the mango (cf. fol. 42r), which forms the basis of more Mughalised renderings both here and in the *Ḥamza-nāma* (fols. 32r and 35r; Pl. 7).

Fol. 66r

The prince's ordeal continues: he is ordered away to be executed for the fifth time.

An ascription on the lower framing line reads *g͟hulām ʿalī,* (by) G͟hulām ʿAlī. Cf. fol. 92r which is by the same artist.

Works ascribed to G͟hulām ʿAlī are quite scarce, being found, as far as we know, in only one other manuscript, the Jaipur *Razm-nāma,* where Hendley misread the name as G͟hulām Nabī. He works there on five miniatures, as a colourist to Basāvana on three (Hendley, Pls. 7, 64, and 65), and to Kānhā on two (*ibid.,* Pls. 29 and 30).

Like fols. 20r, 58r, and 63v, the painting is derived from the *Candāyana* group. The flat patch of red colour, however, is a survival from the Western Indian school or the *Caurapañcāśikā,* group, and is notably absent from future Mughal work. Mughal features, nevertheless, are quite strong. The streaked clouds are beginning to obliterate the curved horizon (cf. Pl. 26), and there is movement and a keen sense of observation in the two guards who hustle the prince away, menacing him with their mace and staff, while he raises his hands in a gesture both of protecting himself and appealing to the king, his father. The symbolic statement is abandoned in favour of a more literal one. In the representation of faces also, we get a heavy type with large eyes, the pupil placed in the centre (the king, the handmaiden, and the guard with the mace on the left), derived from pre-Akbar idioms, and another in which the face is in more naturalistic proportion, the smaller eyes shown in profile and the pupils placed in the direction of the glance (the vizier behind the handmaiden, the guard flourishing a staff). The sword and *jāmā* of the fly-whisk bearer are allowed to project beyond the margin.

Fol. 67r
The deceitful wife returns to her terrace after caressing her lover.

An ascription on the left margin reads *srawan,* (by) Śravaṇa.

The painter, whose name can also be read as Sarwan, is ascribed four paintings in the British Museum *Dārāb-nāma* (fols. 37r, 47r, 47v, and 80v), in three of which he works unaided. Otherwise he is only known as a colourist in the Patna *Tārīkh* (fol. 136v), the Jaipur *Razm-nāma* (Hendley, Pls. 54, 69, 70, and 46), and the Victoria and Albert *Akbar-nāma* (Nos. 18, 45, 66, 88, and 114). The latest miniature ascribed to him is in the National Museum *Bābar-nāma,* fol. 78. This would tend to indicate that Śravaṇa did not develop into a master entrusted with the work of a designer in joint compositions.

The miniature, both in its treatment of space and colour, is considerably Mughalised as are the comparatively delicately painted figures. About the only archaistic vestige is the stiff scarf with projecting tassels.

Cf. fols. 124r and 127r, by the same artist, but with stronger pre-Akbar features.

Fol. 67v
The deceitful wife ejects the procuress after blackening her face.

Attributed to Śravaṇa, the miniature being by the same artist as fol. 67r.

Stronger vestiges of the *Candāyana* style are present in the striped dome, bird-shaped bracket (Pl. 109), and streaked yellow bricks. The light ground with dots and golden check marks, as well as the deep pink floral pattern of the pillow and the canopy are from the same tradition. The female figures are very similar to some of those found in the *Ḥamza-nāma* in general treatment, dress and jewellery, except that the scarves are still quite angular here (Pls. 2 and 4).

Fol. 68r
The farmer, father of the son with the deceitful wife, steals away with her anklet while she is in bed with her lover.

Attributed to Painter F.

The ascription on the left margin is illegible.

The miniature is very similar in style to fols. 67r and 67v, but seems to be by another painter. The pattern of the bed cover and the tasselled fringe, also seen in fol. 63v and 70v, are reminiscent of the *Caurapañcāśikā* group.

Fol. 68v

The deceitful wife persuades her husband to sleep in the same place where she had previously slept with her lover.

Attributed to Painter F. Cf. fol. 68r by the same artist.

Vestiges of the *Candāyana* tradition are to be noticed in the building to the right, with striped dome and blue and gold brick pattern, the sky with waving lines indicating clouds (Pl. 111), and the deep pink floral design of the canopy.

Fol. 69v

The prince sent back to the place of execution for the sixth time.

Attributed to Sūrajū. Cf. fols. 72r, 73r, and 79r by the same artist.

Traces of a cut off ascription on the left margin.

Fol. 70r

The merchant's clerk replaces the sugar purchased by the philandering wife with gravel.

Attributed to Painter G. Cf. fol. 70v by the same artist.

Indo-Persian features are noticeable in the pale ground and the blue sky with arabesque clouds in gold (Pl. 105). *Caurapañcāśikā* traditions are also to be seen in the bed with its chequer design, the vessels beneath the bed, and the peculiar green spots on the panelling below (Pl. 82; also fol. 43r). The figure of the woman is also derived from that style, but the men, except for the clerk, are considerably Mughalised.

Fol. 70v

The husband berates his wife for purchasing gravel instead of sugar.

Attributed to Painter G.

An illegible ascription is present on the right margin.

Close similarities exist between this miniature and fol. 70r, indicating that they were done by the same artist. These include the male and female figures, the mauve and maroon chequer patterns on the canopy and on the bed, the pattern on the pillow and the lady's skirt, and the building with identical crenallations decorated with a cross.

The scalloped curving horizon bordering a triangular section of light blue sky, the arabesque clouds, and the mauve background are derived from Indo-Persian and *Candāyana* traditions (Pl. 104; and *New Documents,* Pl. 157). Clearly recognisable

Caurapañcāśikā features are patterns on the cushion and coverlet, its tasselled fringe, and the typical chequer pattern of the canopy; and also, quite strikingly, the heavy figure of the lady with large *paḍol*-shaped eyes, full cheeks, pointed chin, squat neck and ivory-peg earring, pink-striped *paṭakā,* and yellow skirt with circular pattern enclosing foliate stars (cf. fol. 63v). Among Mughal features are the figure of the man with small, modelled face and expansive gestures, and the building to the left, which is derived from structures of a type seen in the *Caurapañcāśikā* tradition (Pl. 76), but which is here endowed with clear depth, thus anticipating the more complex and spatially consistent arrangements of the *Ḥamza-nāma* (Glück, Taf. 45). The chequer pattern of the canopy also occurs in that manuscript (Pl. 15), as does the striped and lozenge pattern of the flounce (Pl. 8).

The composition of the scene is considerably simplified in the Beatty *Ṭūṭī-nāma* (Pl. 53), the painter confining himself to the bare essentials, namely the couple conversing with each other. The retention of the mauve and maroon chequer pattern of the canopy of the Cleveland *Ṭūṭī-nāma* in the Beatty miniature – the motif being quite rare in that manuscript – is significant, for it indicates the distinct possibility of its painter adapting his work from the Cleveland version.

Fol. 71v

The handmaiden again pleads for the death of the prince.

There is a partially severed ascription on the left margin which reads *lā* [*lū*], (by) Lālū.

The composition is fairly typical of a Mughal court scene. The king is on his throne attended by standing and seated courtiers, diagonally placed. The young prince, his unwound turban piled loosely on his head, stands between the vizier and the wicked handmaiden. The colouring is rich, as in the *Ḥamza-nāma,* no traces of the pale tonalities of the *Candāyana* group being present.

Fol. 72r

The young prince recounts his experiences to his father, the king.

Attributed to Sūrajū. Cf. fols. 69v, 73v, and 79r by the same artist.

The miniature is similar in composition and style to fol. 71v, which is by Lālū, but it does not seem to be by the same artist. The handmaiden is shown standing in the arched entrance to a room. Her figure, in its rhythmical contours, ornaments, and the projecting scarf with stiffly standing tassels, is evocative of the *Caurapañcāśikā* group (Pl. 84).

Fol. 73r

The young prince is crowned and the wicked handmaiden is executed.

An ascription on the column separating two lines of the text reads *surjū,* (by) Sūrajū. Cf. fols. 69v, 72r, and 79r by the same artist.

One miniature entirely painted by Sūrajū occurs in the Patna *Tārīkh,* fol. 254. We also know of a painter named Sūrajū Gujarātī in the same manuscript (fol. 61v), but it is not clear if he is the same person.

Drums are beaten and trumpets blown as the prince sits on the throne. The body of the handmaiden with severed head lies on a panel of tiled floor in the foreground, while the king, who has relinquished the throne to his son so that he might take up the life of an ascetic, is walking away from the city walls, holding a staff and a rosary.

Continuous narration, a pre-Mughal technique, is seen here. Parallels with the *Ḥamza-nāma* are noticed in the soft grassy ground, the tree with yellow-lined leaves diminishing towards the edges, and in the pink, softly shaded rocks (Pls. 31 and 11).

Ninth night

Fol. 75v

The parrot brings a fruit from the Tree of Life to the king of Syria.

An ascription on the lower margin reads *lālū,* (by) Lālū. For other works with the same ascription see fols. 71v, 81v, 83v, and 84v. Fols. 78r and 84r can also be attributed to him on grounds of style.

The consistency between style and ascription in these several paintings suggests the general authenticity and correctness of the ascriptions in this manuscript. This is further confirmed by the fact that Lālū is an almost unknown painter of the Akbar period, and we only know of one other work ascribed to him. It occurs in the Jaipur *Razm-nāma* (Hendley, Pl. 16) where he is working as a colourist to Lāla. The ascription was misread by Hendley as "Qālu," so that the painter, for all practical purposes, was hitherto unknown.

Fol. 78r

The old man eats of the fruit of the Tree of Life, but drops dead.

Attributed to Lālū.

The miniature is close in style to fol. 75v in composition, colour, and figural types. Vestiges of the *Candāyana* style can be seen in the shape and decoration of the columns and the fine line of the lightly brushed arabesque on the yellow band at the top (Pl. 110).

The Beatty *Ṭūṭī-nāma* miniature of the same scene (Pl. 54) is very similar in composition and is obviously an adaptation of the Cleveland example. The figures are more refined and expressive, as is the parrot which flutters its wings at the commotion. The colours are cooler and vestiges of the *Candāyana* style, notably the thin arabesque on yellow, are absent. The faces too are much less heavy, being quite removed from the *Ḥamza-nāma* style in contrast to the Cleveland miniature which is much closer to it.

Fol. 79r

The king plucks fruit from the Tree of Life with his own hands and feeds it to a lady.

Attributed to Sūrajū.

An ascription on the left margin reads [*su*]*rjū,* (by) Sūrajū.

Cf. fol. 72r, the female figures being particularly similar. The general type is also seen in the *Ḥamza-nāma* (Pl. 4) where the representation is more Mughalised.

Tenth night

Fol. 80v

The parrot addresses Khujasta at the beginning of the tenth night.

An ascription on the right margin reads *iqbāl,* (by) Iqbāl.

Ascribed works to Iqbāl are very few. One of them is in the Jaipur *Razm-nāma,* where he works as colourist to Khemakarana (Hendley, Pl. 20), and the other is in the British Museum *Dārāb-nāma,* where he is responsible for one painting (fol. 10r).

Memories of the *Caurapañcāśikā* group survive in the large face and the treatment of the scarf. Otherwise, the miniature is very close to the *Ḥamza-nāma* with its blue carpet bordered in red and decorated with floral arabesque, the panelled eaves with floral design on the architrave, the hexagonal pavilion with domes and details of architectural ornamentation (cf. Glück, Taf. 27; *Ḥamza,* Pl. 33; Pl. 15). The decoration, however, is not as accomplished and elegant, and the sure relationship between spaces that is generally present in the *Ḥamza-nāma* is here absent.

Fol. 81v

The vizier's son receives the magic wooden parrot from the wife of the merchant's son, who is drunk, and has a replica made by a carpenter.

An ascription on the left margin reads *lālū,* (by) Lālū.

The technique of narration employed is similar to that in fol. 73r, three incidents separated in time being shown in the same miniature. The female figure is very close to those in the *Ḥamza-nāma* (Pl. 2).

Fol. 83v

The vizier's wife sends the magic wooden parrot to her lover, the monk, who exchanges it for an ordinary one.

An ascription on the right margin reads *lālū,* (by) Lālū.

The golden mound is heavily shaded with pink towards the top to give it the shape of a hill. The scarf of the lady is of the *Caurapañcāśikā* variety.

Fol. 84r

The monk returns the magic parrot to its rightful owner, the merchant's son.

Attributed to Lālū.

There is no ascription, but the miniature is obviously by the same hand as fol. 83v, which is ascribed to Lālū.

Fol. 84v

The magic parrot of the merchant's son talks to the vizier's son.

An ascription on the left margin reads *lālū,* (by) Lālū.

The style is identical to fols. 75v and 84r which serves to confirm the correctness of the ascriptions.

Eleventh night

Fol. 87r

The parrot addresses K͟hujasta at the beginning of the eleventh night.

Attributed to Banavārī 2.

An ascription on the lower margin reads *banwārī,* (by) Banavārī.

The same ascription is seen on fol. 50v, but the two miniatures appear to be by different hands. One of the two ascriptions is either incorrect or two different persons are indicated. Both a Banavārī Kalān and a Banavārī K͟hurd are known as artists of the Akbar period and this could possibly be Banavārī K͟hurd (see note under fol. 50v).

Night is indicated by the black ground and crescent moon. The lady carries memories of the *Caurapañcāśikā* group. Otherwise, the work, though somewhat coarse, is indistinguishable from the *Ḥamza-nāma* style.

Fol. 89r

The Brahman is asked by the Rājā to bring the king of the Ocean to his nuptial feast on pain of death. The Brahman's predicament is conveyed by the wind to the fish who carries the news to the king of the Ocean.

Attributed to Painter H. Cf. fol. 89v by the same artist. He is among the outstanding painters of the manuscript.

The Brahman does not stir out of his house in the text, but here he is shown standing next to the ocean, leaning forward expectantly with a string of beads in one hand and an offering in the other. The swirling water, which fills the picture except for a corner, is splendidly and imaginatively painted and throws up heavy foam on the shore, probably an allusion to the king of the Ocean quivering and foaming at the mouth upon hearing the preposterous request.

Similar depiction of water, both the tumultuous movement and the foaming edges, are to be found in the *Hamza-nāma* (Pl. 1; Glück, Taf. 7; *Hamza,* Pl. 13), but the rendering here is specially majestic.

Fol. 89v

The creatures of the sea are asked by the king of the Ocean to take a message to the Brahman.

Attributed to Painter H. Cf. fol. 89r by the same artist.

The text says that these creatures were the sea dragon (who in winter hunches the middle part of his body to better receive the rays of the sun), the whale, the turtle, the crab, the crocodile, the alligator, and the frog. The sea dragon, like an "arc in the sky," is faithfully depicted as are the other creatures. The crocodile and the alligator are both partly fanciful. I suppose the one with the longer snout is meant to be the crocodile.

Fol. 92r

The king of the Ocean assuming human form betakes himself to the Brahman's house to be taken to the Rājā.

Attributed to Ghulām ʿAlī.

The miniature is very similar to fol. 66r and seems to be by the same artist.

The king of the Ocean, wearing a *dhotī,* scarf and crown, is greeted by the Rājā. The patch of red is a survival from the Western Indian style and the *Caurapañcāśikā* group. Mughal painting abandons its use, but it continues to be popular in the schools of Rajasthan. The plasticity and easy-flowing manner in which the ends of the scarf are shown are Mughal features. Notice also the turban with a small *kulāh,* but without the trellis, of a type that is often seen on the heads of Hindu priests in Akbar period painting (cf. fol. 89r).

Fol. 93v

The Brahman, unable to select from the four gifts of the king of the Ocean seeks the Rājā's advice.

Except for the absence of the red background, the miniature resembles fol. 92r. The light blue wall and the yellow brick-tile floor with black streaks are *Candāyana* features. The Rājā's throne is decorated with the oft repeated mauve and maroon chequer pattern.

Twelfth night

Fol. 95r
The parrot addresses Khujasta at the beginning of the twelfth night.

Attributed to Tārā 1. Cf. fols. 60v, and 102v which are by the same artist.

The miniature, though more archaistic, is strongly reminiscent of fol. 60v, notably in the rendering of the female figures. Notice particularly the sense of rhythmical movement and the carefully rendered gauzy texture of the *oḍhanī,* which are characteristic of the artist.

Fol. 97r
The street cleaner, on his way to meet king Bhojarāja, sleeps under a tree where four thieves disguised as fellow travellers deprive him of a priceless pearl.

Attributed to Painter I. Cf. fol. 97v which is by the same artist.

The helplessly asleep figure, whose pouch is being stealthily removed by a thief, is convincingly rendered. The banyan tree, its trunk encrusted with a network of aerial roots, and foliage consisting of green-veined leaves with yellow rims and freshly sprouted pink sprigs (cf. Pl. 34), is naturalistically done. It is inhabited by large white birds, though they, as well as the tree, do not possess the accomplished ease of the *Ḥamza-nāma* (cf. Glück, Abb. 38; *Hamza,* Pl. 14). Further affinities are evident in the diagonally arranged figures of the three seated men to the left, the sprawling street cleaner, and the thief with out-stretched arms, all of which contribute to pictorial depth.

Fol. 97v
King Bhojarāja tries in vain to ascertain the whereabouts of the pearl from the four travelling companion.

Attributed to Painter I. Cf. fol. 97r by the same artist, the figural types being almost identical.

The king wears a trellised *kulāhdār* turban of exactly the same type as seen in the *Candāyana* and *Caurapañcāśikā* groups (Pls. 107 and 82), though the *kulāh* (cap) does not have as prominent a projection. This item of dress, among other features, is a rare survival from pre-Akbar painting, and was soon abandoned, though a modified version, without the trellis, is fairly common in paintings of the Akbar period, being generally worn by Hindu priestly types.

Fol. 98v

The daughter of the merchant of Mazanderan asks the gardener for the rose.

Attributed to Painter J. Cf.fols. 99v and 100v which are by the same artist, who is among the most conservative in the entire manuscript, and was originally trained in the style of the *Caurapañcāśikā* group.

The gardener, holding a shovel, faces the lady who is accompanied by three attendants one of them holding a basket of flowers in an upraised arm.

The painter holds on to his pre-Mughal traditions tenaciously. Features of the *Caurapañcāśikā* group are particularly clear in the sharply projecting *oḍhanī,* the ends edged with stiff tassels, the textile patterns, and the stylised gestures. The manner of depicting the breasts as intersecting three-quarter circular forms; the ornaments, noticeably the collar necklace and ivory pegs in the ears; the sharply projecting *paṭakās* and the rhythmical outlines are also shared with the *Caurapañcāśikā* group (Pls. 80 and 84). Nevertheless features developed in the *Ḥamzanāma,* though in a rudimentary stage, are also present, notably in the less unremittingly rigid line, the attempt to model the face and the garments, and the yellowish tufts of grass.

Fol. 99v

The merchant's daughter encounters a wolf and bandits on her way to meet the gardener in order to keep her promise.

Attributed to Painter J. Cf. fols. 98v and 100r which are by the same artist.

The landscape, here as in fols. 98v and 100r, is different from what we get in the other miniatures in this manuscript. The foliage is essentially ornamental, being neither feathery, stippled, or of the naturalistic type. True, the tree trunks are softly shaded, but the treatment of the leaves is based on the *Caurapañcāśikā* tradition. Tree-tops are thus ovaloid in shape, the individual leaves being massed between the branches with fresh sprigs represented in shades of pink, or disposed in horizontal garland-like rows (Pls. 73 and 80; and *New Documents,* Col. Pl. 19b). In some of the trees, notably the third from the right, we see circular flowers all along the edge; these are vestiges of a more ornamental convention of the *Caurapañcāśikā* group (Pls. 80 and 81).

This type of decorative rendering of trees, inherited from Indian pre-Akbar traditions did not find favour at the Mughal atelier and was soon given up. Even in this manuscript, its occurrence is quite rare. The schools of Rajasthan, on the other hand, which do not display the naturalistic preferences of the Mughal school, continued these traditions. A good example is the Chawand *Rāgamālā* supposedly dated 1605, from Mewar (Pl. 94). Resemblances between paintings of this type and the Cleveland *Ṭūṭī-nāma* are explicable in terms of both borrowing from a common source, and in no way suggest a late date for the Cleveland miniature. Rather, a posterior date for the Chawand *Rāgamālā* is indicated by the presence of Mughal influences such as the softly daubed foliage of two of the trees, which not only recalls work in the *Ṭūṭī-nāma,* but also the *Ḥamza-nāma* (Pl. 20), particularly as this device does not occur in pre-Mughal painting.

fol. 100v
The merchant's daughter meets the gardener.

Attributed to Painter J. Cf. fols. 98v and 99v which are by the same artist.

The shaded pink brick pattern at the base, meant to designate a paved path, is not normally found in the *Caurapañcāśikā* group; rather it is a frequent feature of the *Candāyana* group. The female figures, both in this and the other two miniatures by this artist, form the basis of more refined versions seen in the *Ḥamza-nāma* (Pl. 2).

Thirteenth night

Fol. 102v
The parrot addresses Khujasta at the beginning of the twelfth night.

Attributed to Tārā 1.

Cf. fols. 60v and 95r which are by the same artist. The identical treatment of the two female figures with their gauzy *oḍhanī,* dress and jewellery, including strings of pearls in the hair, so close to the *Ḥamza-nāma* (Pl. 2), is noteworthy. The bolster on the bed, decorated with a border of flowers that do not quite touch it, bears a striking resemblance to a device commonly found in the *Caurapañcāśikā* group (Pl. 81) and in another miniature of this manuscript (fol. 282v).

Fol. 105r
The infant son of the king of Isfahan responds to music.

Attributed to Painter K. Cf. fol. 125v which is by the same artist.

Of the three infants in cradles, only the one in the centre listens attentively while the others sleep on unaffected. The figures below are arranged in a circle around a wise man, the lower part consisting of musicians playing on a variety of instruments.

The pale mauve colour of the top panel and the arabesque of the large carpet derive from the Indo-Persian and *Candāyana* groups as do the noticeably archaistic faces (Pls. 101 and 108). The circular arrangement of the figures is hardly ever seen in pre-Akbar painting, but is a common enough device in the *Ḥamza-nāma* and later (Glück, Abb. 10). The heads are also strikingly similar to a detail in the *Ḥamza-nāma* (Pl. 33) which was probably done by the same artist. It may be noted here that resemblances between the Cleveland *Ṭūṭī-nāma* and the *Ḥamza-nāma* are the strongest when the *Ḥamza-nāma* painter is working on a scale comparable to the much smaller miniatures of the *Ṭūṭī-nāma.*

Fourteenth night

Fol. 108v

The invention of musical instruments from the intestines of a monkey.

From the music produced by the wind blowing through the intestines were developed the *rebāb* and the *vīṇā* (harp of Nakhshabī) played by the musicians shown in the miniature. The drawing of the trees is cursive, but the posture of the *vīṇā* player is most expressive.

The plain green background is stippled with yellowish dots at the edge. This technique is rare in this manuscript, but it occurs in a few paintings of the *Ḥamza-nāma* (Glück, Taf. 1 and 44; *Ḥamza,* Pls. 1 and 54) and other contemporary works (Pls. 66 and 44).

Fol. 110v

The origin of music from a fabulous bird of India which had seven holes in its beak.

A young man with delicately drawn features, holding a *vīṇā* in one hand and a book in the other, is shown seated beneath a tasselled canopy strung between two trees. His quiver, sword, and shield hang from a tree to the left, its finely stippled foliage recalling in some measure the work of Basāvana and Dasavanta (cf. fols. 35r and 32r). While an attribution to Basāvana cannot be altogether excluded in view of the wet brush, the soft line, the voluminous rocks and the exquisitely painted birds, the more formal rendering suggests the work of some other master. The seated male figure, in particular, is somewhat removed from Basāvana's style (cf. the seated king in fol. 36v). In a certain sense the work is more reminiscent of Dasavanta, but lacks his freedom, the brushwork being more precise, best exemplified in the handling of the feathery foliage of the tree to the left (cf. fol. 32r). One is also reminded of the style of Painter U, but the technique here is more assuredly refined, and we do not think it is by him.

Characteristics derived from the *Caurapañcāśikā* are apparent not only in the ornamental foliage of the mango, but also in the red background and the general composition which anticipates the kind of arrangement seen in later *Rāgamālā* miniatures from Rajasthan. Here it could be derived from a *Rāgamālā* of the

Caurapañcāśikā tradition, such as the *Rāginī Bhairavī* in the Victoria and Albert Museum (Archer, *Central Indian Painting,* Pl. 3, p. 8),

Fifteenth night

Fol. 112r

The parrot addresses Khujasta at the beginning of the fifteenth night.

Profuse use of floral arabesques seen here on the red wall behind the lady, on the yellow bed sheet, on the blue wall behind the bed where it is delicately painted in gold, on the yellow rug at the edge of which are stands holding water pots one placed on top of the other, together with the floral patterns on the bolster and the canopy, are all strongly reminiscent of the *Candāyana* and Indo-Persian groups (Pl. 101; and *New Documents,* Pl. 161). The colouring is also derived from the *Candāyana* group, except for the wall behind the lady which is red. The lady, as is so often the case, has affinities to the *Caurapañcāśikā* group, which is apparent in the bold quatrefoil ornament on the skirt, and the angular disposition of the *paṭakā,* narrow near the waist and projecting sharply below the knees.

Fol. 113r

The lion disturbed by mice who eat the food trapped in his aging teeth.

Attributed to Painter L. The style is quite Mughal, here as in fols. 114r and 115v which are by the same artist.

The paint has chipped off to the left of the rock, revealing the stylised foliage of a tree, which was painted over and replaced by the present tree which is in the *Ḥamza-nāma* style (Pl. 21).

A tiger, instead of a lion (one animal is often shown for the other, the same Persian word denoting both), is shown in a clearing amidst rocks. The mice scamper in from all directions.

The Beatty *Ṭūṭī-nāma* miniature (Pl. 55) representing the same theme shows many advances, notably in the depiction of the lion (not a tiger) which is lying on its side, helplessly asleep, head foreshortened. The tiger of the Cleveland miniature, on the other hand, is a Mughal version of an Indo-Persian animal (cf. fol. 150r and Pl. 103). The colour, the rocks, and the tree are also close to the *Ḥamza-nāma,* while the Beatty miniature relates to a later phase of Akbar period painting.

Fol. 114r

The wolf advises the lion to consult the cat.

Attributed to Painter L. Cf. fols. 113r and 115v which are by the same artist.

Fol. 115v

The cat attacks the mice which disturb the lion.

Attributed to Painter L. Cf. fols. 113r and 114r which are by the same artist.

Sixteenth night

Fol. 119v

The daughter-in-law of the king of Banaras, charmed by the music of a vagabond, comes down to meet him.

With the *Candāyana* style may be compared the pale colours of the lower panel and the lightly drawn arabesques of the upper panel; while the red ground of the upper panel, the female figures, and the ornamental patterns of the bed cover and its fringe of tasselled lappets point to the *Caurapañcāśikā* group.

Cf. fols. 59v and 69v.

Fol. 120r

The vagabond crosses a stream with the possessions of the daughter-in-law of the king of Banaras and absconds.

The Beatty *Ṯūṯī-nāma* representation of this incident (Pl. 56) is based upon this miniature, as is evident in the composition with its diagonally arranged stream, and the lady and her lover on opposite banks. Instead of the basket-pattern, the water is done in swirls, the brightly coloured boulders edging the stream are omitted, and the soft green and yellow mounds of grass are replaced by stippled dots. The single tree is now placed on the hither side of the stream in such a way as to lend depth to the composition. It is interesting to note that the feathery foliage of the Cleveland example is retained, though used in a different type of tree.

Fol. 121r

The daughter-in-law of the king of Banaras sees the jackal deprived of its food by a bird, as it unsuccessfully attempts to catch a fish.

Similar in style to fol. 120r, and perhaps by the same artist. The rocks closely resemble the *Ḥamza-nāma*, being gently highlighted with washes of white (Pl. 12). The tree with its closely packed leaves, is derived from the *Candāyana* tradition, but is also found in the *Ḥamza-nāma* (Glück, Taf. 21).

Fol. 122v

The daughter-in-law returns from her misadventure, feigning insanity.

The miniature is very close in conception to fols. 66r and 92r. It seems to be basically derived from the *Candāyana* group, but the strong red background and the female figures in the kiosk are reminscent of the *Caurapañcāśikā* style and earlier (Pl. 68).

Seventeenth night

Fol. 124r

The parrot addresses Khujasta at the beginning of the seventeenth night.

Attributed to Śravaṇa.

The room with its pelmet-like projections from the ceiling is derived from the *Candāyana* tradition (Pl. 112). Other features from the same group are the shaded bricks, the pale tones of colour, and the flat composition. The heavy green curtain is softly modelled, very much in the *Ḥamza-nāma* manner.

Pre-Akbar traditions are much more evident in this miniature and fol. 127r than they are in fols. 67r and 67v which are also by Śravaṇa. The female figures, so close to the *Ḥamza-nāma* (cf. Pl. 4), however, are identical, and form the basis of the attribution.

Fol. 125v

The merchant Manṣūr departs on a sea-voyage leaving his wife behind.

Attributed to Painter K. Cf. fol. 105r which is by the same artist.

The double storeyed pavilion with striped dome and tiled wall, the pattern on the bed and bolster, and the boat with ornamentally rendered plank construction are *Candāyana* features, as is the basket-patterned water with only the most gentle hint of foam where it laps the boat. The pale hill, with tufts of grass, projecting into the sky is derived from the Indo-Persian tradition (Pl. 99). The figures in the boat are identical with those of fol. 105r, a miniature equally dependent on the *Candāyana* and Indo-Persian traditions.

Fol. 127r

The old procuress conveys the young man's message of love to Manṣūr's wife.

Attributed to Śravaṇa. Cf.fol. 124r which is by the same artist.

The composition is similar. The tall arched panels on either side occur both in the *Candāyana* and Indo-Persian traditions (Pl. 107), as do the bird bracket and the hexagonal patterns on top (Pls. 109 and 112).

Fol. 128r

The young man takes leave of the monk who teaches him a magic formula.

Attributed to Painter M. Cf. fols. 128v and 129v which are by the same artist.

The miniature shows striking similarities to a painting in the Lalbhai *Khamsa* (cf. Pl. 105), which belongs to the Indo-Persian group. The striped domes and the tiled decoration are derived from *Candāyana* traditions.

The central dome was originally semi-circular, but has been rather crudely corrected to an onion shape to correspond roughly to types found in the *Ḥamza-nāma* (Pl. 32).

Fol. 128v

The young man changes himself to look like Manṣūr, and thus inveigles himself into the bed of Manṣūr's wife, but is put off by her.

Attributed to Painter M. Cf. fol. 129v which is by the same artist.

The figures of the bearded men and the style of the arabesque on the wall are identical in this miniature and fol. 128r, though the colouring here is richer.

Fol. 129v

The false Manṣūr punished before the judge and expelled from the city.

Attributed to Painter M.

Of the three miniatures by this artist, this one possesses the strongest Mughal features. The figure of the lady and of the attendants are similar to fols. 128r and 128v, suggesting that they are by the same artist.

Eighteenth night

Fol. 130v

The parrot addresses Khujasta at the beginning of the eighteenth night.

Among *Caurapañcāśikā* conventions are the patterns on the skirt of the lady and the bed cover, as well as the knotted brackets (Pl. 82), both of which continue to be found in the *Ḥamza-nāma* (Pls. 19 and 34).

Fol. 131v

The prince meets a carefree dancing *darwīsh* whose good fortune he purchases for his ring.

The heavy face of the prince is comparable to those on fols. 66r and 92r.

Fol. 132r

Nīkfāl, the fortune of the prince in the form of a woman, offers to accompany him.

Attributed to Painter N. Cf. fol. 132v which is by the same artist.

The figures may be contrasted with those of fol. 131v, being more delicately drawn, the head done in a more naturalistic proportion to the body. Clumps of yellowish grass cover the earth, the forest being symbolically indicated by the presence of the deer. A finely drawn arabesque is sketched across the blue sky. Beyond the curving horizon are the partially visible fronds of a palm tree. The miniature recalls paintings of the Indo-Persian group (Pl. 103).

Fol. 132v

The prince, having deprived the snake of its natural food, a frog, feeds it with a piece of his own flesh.

Attributed to Painter N. Cf. fol. 132r which is by the same artist.

The mango tree shows an attempt to transcend the decorative limits set for it, but is still not as naturalistic as that of fol. 110v, and fols. 32r and 35r. The pool, of a basically Persian type, occurs frequently in the *Ḥamza-nāma* (Glück, Abb. 31). The female figure, with a distinctive blouse pattern derived from the *Caurapañcāśikā* group (Pl. 83), is very close to types in the *Ḥamza-nāma* (cf. Pl. 4). Notice also the use of continuous narration, the release of the frog and the feeding of the snake being shown in the same picture.

Fol. 134v

The prince and Nīkfāl are joined by Kh̲āliṣ and Muk̲h̲liṣ who are the grateful snake and frog in human form.

An attempt is made to merge the foreground, decorated with yellowish flowering plants, into the pale mauve background with its stylised tufts of grass. Gold sky with waving golden lines. The heavy faces are reminiscent of fols. 66r and 92r.

Fol. 135v

The prince, with the help of Muk̲h̲liṣ who changes into a frog, recovers the ring lost in the sea, and returns it to the king.

The sea, more like a lake with embankments, has a boat with curved prow in the shape of a bird's neck and head. The water is a modified basket-pattern like that also seen in the *Ḥamza-nāma* (Pl. 10; Glück, Taf. 11 and 13; *Ḥamza,* Pls. 17 and 19).

Cf. fol. 97v which is quite similar in style.

Fol. 136r

Kh̲āliṣ repays the prince for his kindness by changing into a snake and sucking the poison from the king's daughter.

The massive pots, covered by a thin soft cloth, may be contrasted with those in fols. 112r and 124r, which are of an archaistic aspect, and compared with those in the *Ḥamza-nāma* (Glück, Abb. 28; *Ḥamza,* Pl. 49).

Nineteenth night

Fol. 138r
The parrot addresses Khujasta at the beginning of the nineteenth night.

Two of the domes are tiled, a Mughal feature. The green floor with gold stripes, the mauve ground, and the lightly sketched arabesque on the yellow architrave are *Candāyana* vestiges.

Fol. 140r
The Brahman's wife who killed a peacock and ate its gall bladder on the physician's advice.

Among features of the *Candāyana* group is the pillar with bird bracket (Pl. 109).

Twentieth night

Fol. 143r
The parrot addresses Khujasta at the beginning of the twentieth night.

Attributed to Painter O. Cf. fols. 160r, 161r, 161v, 168v, and 170v which can also be attributed to him on grounds of style. The work is similar to that of Painter P.

Archaistic traits, except for the ivory-peg earring, are almost absent from the figure of the lady, though they are present in the colour of the pale mauve wall and the stand with piled pots (Pl. 109). Notice the peculiar rounded head, to be distinguished from the more usual type which is squarish in shape.

Fol. 144v
The three suitors fight among themselves for the hand of the devotee's daughter.

Attributed to Painter P. Cf. fol. 147r which is by the same artist.

The style is similar to that of Painter O (cf. fol. 143r), but is by a different artist. Miniaturistic proportions with the heads in a naturalistic relationship to the body are to be seen in the figures fighting in the foreground, and they are strongly reminiscent of the *Ḥamza-nāma* (cf. Pls. 8 and 33). It is the same scale as used by Dasavanta (fol. 32v), but the workmanship is of much lesser quality.

Fol. 146r
The suitors take the devotee's daughter out of her tomb after breaking it open when the physician discovers she is still alive.

Attributed to Painter Q.

The miniature can be compared to fol. 62v, showing a similar movement away from the *Candāyana* and towards the Mughal idiom. The cluster of shaded rocks in the corner from which cascades a stream of water into a pool banked by lush flowering foliage (cf. Pl. 10); the proportionate figures in the foreground; and the sky with soft wispy clouds instead of arabesques (cf. Pl. 26) demonstrate a close relationship with the *Ḥamza-nāma*, as do the figures (cf. Pls. 8 and 33).

Fol. 146v

The third suitor strikes the devotee's daughter and thus restores her to life.

Attributed to Painter Q. Cf. fol. 146r which is by the same artist, the male figures being quite identical. The arabesque cloud of Indo-Persian and *Candāyana* derivation (Pls. 99 and 106), avoided in that painting, is here present.

Fol. 147r

The three suitors again begin to quarrel among themselves for the hand of the devotee's daughter.

Atrributed to Painter P. Cf. fol. 144v which is by the same artist.

Twenty-first night

Fol. 150r

The Brahman comes upon a lion who has a deer and a gazelle as his viziers.

The miniature adheres to the *Candāyana* tradition to a greater degree than fol. 146 in the pronouncedly pale tones of colour, the curving horizon, the pink ground with sparse tufts of grass, the leaning tree with its closely packed leaves edged in gold, the basket-patterned water, and the foliate verdure of the banks which resembles the leaves of the tree (cf. Pl. 111).

Movement towards the Mughal style is to be seen in the rocky shapes painted at the horizon's edge, the shaded boulders placed diagonally in the bottom right corner, the *dhotī* of the Brahman, and the rendering of the animals, particularly in the fairly delicate brushwork of the lion's mane.

Fol. 152r

The wolf and the jackal, serving as viziers, instigate the lion who pursues the Brahman up a tree.

The miniature provides an instructive comparison with fol. 150r, representing one further step in the evolution towards the Mughal style. The composition is basically the same, but the introduction of the palm tree and the cypress, as well as the extension of the main tree beyond the horizon, add to the depth. The colour, too, is similar to fol. 150r, but the tones are deeper. The lively fluttering foliage with the leaves becoming smaller and fainter towards the edge (cf. Pl. 21); the basket-pattern of the water beginning to break up into swirls; the shrubs on the banks (cf. Pl. 10); the tall swaying grass and reeds (Pl. 12); the softly modelled rocks; and the expressive drawing of the animals are closer to the *Ḥamza-nāma* style in this miniature than in fol. 150r.

The golden arabesque in the sky in an archaistic survival as is the square face, which, however, continues to be seen in the *Ḥamza-nāma* (Pls. 16 and 27).

Twenty-second night

Fol. 154r

The parrot addresses K͟hujasta at the beginning of the twenty-second night.

Attributed to Painter R. Cf. fol. 179v which is by the same artist.

In spite of the curved horizon, and the yellow brick tiles at the base, the heavy colour and tiled domes give a markedly Mughal aspect to this miniature.

The miniature may be compared with a similar painting in the Beatty *Ṭūṭī-nāma* (Pl. 57). Both pre-Akbar and *Ḥamza-nāma* features are absent there. The colours are harmonious; the use of arabesque and tile patterns much less conspicuous; and the female figures more fully modelled and without any traces of the conspicuous angularity that is to be seen in the Cleveland miniature and to some extent in the *Ḥamza-nāma.*

Fol. 155r

The court jester meets a Zangī dancing with joy, and learns from him that the cause of his happiness is his assignation with a woman who is the jester's own wife.

The ugly Zangī, with open mouth and bare teeth, as well as the court jester, are sensitively painted against a background of pale yellow with sparse tufts of grass.

A narrow strip at the base with a design similar to the deep pink patterns seen elsewhere in this manuscript, but unconnected to the rest of the picture, and shadows of other figures behind the paint indicate that the miniature had been painted over a previous effort. Infra-red photography confirms this, and also reveals a word in Arabic characters at the top, reading *zangī.* Apparently the notation was to remind the painter of the scene he was to depict, but he erred, perhaps because he was unable to read the note, and painted instead what seems to have been a representation of the court of the *amīr* of Kirman, with whom the jester was employed. This error was subsequently corrected.

Cf. fol. 197r where a similar error was committed.

Fol. 158v

As punishment, the jester's wife and the Zangī are thrown into fire and the *amīr's* wife and the mahout are trampled by an elephant.

Attributed to Painter C.

The miniature seems to be by the same artist as fol. 29r; the drawing of the figures, the light green ground with yellowish grass, its tufts obscuring the horizon and the conventional streaks composed of three dots and a dash are almost identical.

Twenty-third night

Fol. 160r

The parrot addresses K̲h̲ujasta at the beginning of the twenty-third night.

Attributed to Painter O. Cf. fol. 143r by the same artist.

A stool is placed below the lady in order to establish her on a support and thus negate the "floating" appearance.

Fol. 161r

The merchant has the hateful skull ground and put into a box.

Attributed to Painter O.

The painting has affinities to the Indo-Persian and *Candāyana* groups in composition, colour, and the delicate ornamention (cf. Pls. 105 and 110). The seated women with round faces are very similar to those in fols. 143r and 160r, and this minature is by the same artist.

Fol. 161v

The merchant's daughter gives birth to a son as a result of eating out of the box. The clever child recognizes the false gems from the true.

Attributed to Painter O.

Two scenes separated in time, the daughter with the box and her son arguing with the jewellers, are shown in the same miniature.

Like fol. 161r, the miniature has affinities to the Indo-Persian group (Pl. 105). The bracketed pillar separating the two scenes is of a type that occurs in the *Ḥamza-nāma* (Glück, Abb. 37), and is popular in the Akbar period.

Fol. 163r

Kāmjūy, the wife of the Rājā, averts her face from the fishes.

The miniature, derived from the Indo-Persian group, is comparable to fols. 161r and 161v. The garden beyond the fence with doorway occurs both in the

Indo-Persian group (Pl. 102) and the *Ḥamza-nāma* (Pl. 32). Patterns on the dresses of the ladies continue to be inspired by the *Caurapañcāśikā* group.

Fol. 164r

The forty wives and their secret paramours being punished by stoning to death.

On both sides are the king's men, hurling stones into a medley of figures consisting of the adulterous wives and their lovers who are dressed as women, the guise which they adopted in the harem.

The victims are piled in a pyramid, firmly imprisoned by their own enmeshed bodies, limbs and arms. The crowded composition, rare in this manuscript, is evocative of the *Ḥamza-nāma,* a relationship which is emphasised by the vehemence of the stone throwers, and the distorted, teeth-baring agony on the faces of the victims (Pl. 30). The vigorous movement of early Mughal painting has been frequently noticed as a distinguishing characteristic. The source of the confused but spirited movement appears to be battle scenes of the type seen in a *Bhāgavata-purāṇa* of the *Caurapañcāśikā* group (Pl. 77).

The triangular composition and the colour with its vivid reds, yellows, blues and whites, and also the green ground, with its grassy tufts, are particularly close to the *Ḥamza-nāma* (cf. Glück, Taf. 31).

Twenty-fourth night

Fol. 165

The parrot addresses K͟hujasta at the beginning of the twenty-fourth night.

The bird cage hangs from an ornamental bracket that is arbitrarily fitted to the framing line, a feature also found in fols. 185r and 198r, and derived from the *Caurapañcāśikā* group (Pl. 78). The chequer pattern on the bed cover is also derived from the same group, but it is painted here in brown and yellow instead of the usual mauve and maroon.

Cf. fol. 160r for a similar miniature.

Fol. 167v

Bashīr confides his love for Ḥabbaẕā to an Arab friend, and sends him to her with a message.

Attributed to Painter S.

Bashīr and the Arab are seen in conversation in the top left corner. Below we see the Arab approaching the cluster of tents which is Ḥabbaẕā's dwelling.

The two separate registers, one with a light blue and the other with a mauve background, are rather effectively united by a curving row of prominent rocks, though the intimate relationship between the various parts of the picture is not as effectively established as in a similar scene from the *Ḥamza-nāma* (Glück, Taf. 30). The colours here are as deep and rich, the tents are similarly painted, and depict the same patterns on the fringes, though the *Ḥamza-nāma* painting is, as is generally the case, richer in decoration and more accomplished in technique (cf. Pl. 8). For similar figural types in the *Ḥamza-nāma* see Pls. 8 and 33.

Fol. 168v

Ḥabbaẕā meets Bashīr under a tree.

Attributed to Painter O. The figure of Ḥabbaẕā strongly resembles that of Khujasta in fol. 143r, and the two paintings seem to be by the same artist.

The main scene, as in the fol. 167v takes place in the top left corner, this de-emphasising of the principle event being a stylistic device often employed by Mughal painters, both in the *Ḥamza-nāma* and later.

Though the compositions of the two miniatures are similar, the painter here has less success in integrating the two planes, the row of rocks at the center being fairly uniform and doing little to unite the upper and lower half of the miniature. The colours too are a little colder, the shading less vivid, so the miniature must have been painted by an artist other than that of fol. 167v.

Fol. 169v

The disguised Arab, substituting for Ḥabbaẕā, is whipped by her husband for refusing a bowl of milk.

Attributed to Painter S.

The miniature resembles fol. 167v both in colour and technique, and is by the same artist. Note the symmetrical composition, the main tent being provided with four small ones at the corners.

Fol. 170v

Ḥabbaẕā's sister, who is sent to console her, discovers the disguised Arab in her place.

Attributed to Painter O.

Though the composition is similar to fol. 169v, the miniature is by the same artist as fol. 168v. The painter has almost achieved the Mughal style here.

Twenty-fifth night

Fol. 171v

The parrot addresses K̲h̲ujasta at the beginning of the twenty-fifth night.

The figure of K̲h̲ujasta is more rhythmically conceived than, for example, in fol. 165v, and to that extent is closer to *Caurapañcāśikā* traditions. This is confirmed by the treatment of the scarf, which is provided with an edge of projecting tassels. These are, however, not black, but are painted in colours that match the skirt.

Fol. 174v

Muk̲h̲tār throws his wife Maimūna into the pit, but she saves herself.

The artist has visualised the pit as a cleft in a hill out of which Maimūna is pulling herself by grasping the branches of a bush. The softly shaded, bulky rocks are highlighted with touches of white (cf. fols. 182r and 183v) and are very similar to representations in the *Ḥamza-nāma* (Pl. 12). This type of treatment is hardly ever to be seen outside the *Ḥamza-nāma,* and becomes scarce even in related works like the Rampur *Astrology* manuscript (cf. Pl. 44).

Fol. 175v

The destitute Muk̲h̲tār meets his wife Maimūna at a holy shrine.

A group of women is present at the shrine, one of them placing garlands and flowers on the grave. In the foreground is the scantily clad Muk̲h̲tār who has been recognised by his wife.

Among parallels to the *Ḥamza-nāma* are the tiled onion dome (Pl. 32), and the slender pillar with herring bone patterns (Pl. 15); while the high curving horizon and the heavy faced female figures with chequered skirts are among the few vestiges of pre-Akbar painting.

Fol. 177r

The lover of Hamnāz, who has been hanged from the gallows, bites off her nose when she kisses him.

Attributed to Painter T. Cf. fol. 177v which is by the same artist.

Blue, shaded mountains of the *Ḥamza-nāma* type have been inadroitly grafted to the mauve ground with curving horizon.

Fol. 177v

In order to falsely implicate her husband, Hamnāz places a knife by his side and lets the blood dripping from her nose stain his clothes.

Attributed to Painter T. Cf. fol. 177r which is by the same artist.

The scale of the figures is small, but the workmanship is relatively coarse.

Twenty-sixth night

Fol. 179v
The parrot addresses Kh̲ujasta at the beginning of the twenty-sixth night.

Attributed to Painter R. Cf. fol. 154r which is by the same artist.

The pavilion with tiled domes is conceived in depth. The birds flying in the sky recall the *Ḥamza-nāma* (Glück, Abb. 21) to which this miniature is even closer in style than fol. 154r.

Fol. 182r
The dethroned frog Sh̲āpūr seeks the help of the serpent.

The curve of the rocks is balanced by that of the tree in a carefully arranged composition. A similar type of tree with several clumps of foliage is to be seen in the *Ḥamza-nāma* (Pl. 3).

Cf. fols. 174v and 183v for a similar rendering of rocks, closely paralleled in the *Ḥamza-nāma* (Pl. 12).

Fol. 183v
The snake enters into an argument with the frog.

The rocky hill is very similar to fols. 174r and 182r. The closest parallels are in the *Ḥamza-nāma* (Pl. 12).

Twenty-seventh night

Fol. 185r
The parrot addresses Kh̲ujasta at the beginning of the twenty-seventh night.

The pillar is decorated with a lotus-petal pattern derived from the *Caurapañcāśikā* group (Pl. 83). The cage hangs from an ornamental bracket as in fols. 165v and 198r.

Twenty-eighth night

Fol. 190v
The king advises the potter of his incompetence for war (Sawyer collection).

The miniature is strikingly reminiscent of the *Ḥamza-nāma*, in the figures (Pl. 8), the flowering plants on the ground (Pl. 10), and the types and decoration of tents (Pl. 8).

Twenty-ninth night

Fol. 192r

The parrot addresses Khujasta at the beginning of the twenty-ninth night.

Here, the cage is suspended on a chain hanging from the border of the miniature, not from the ceiling as in fol. 179v.

Fol. 193r

The monkey, serving as the lion's chamberlain, converses with the lynx and its mate who have arrived with their cubs to settle in the lion's domain.

The text describes the lion's home as a luxurious meadow with flowering tulips and the artist has taken some pains to depict it as such, with plants of bright green and yellow foliage. At the base of the trees are bright rocks (Pls. 21 and 10).

Fols. 193r, 194v, and 196r are quite similar in style, but it is uncertain if they are by the same painter.

Fol. 194v

The lion returns to his territory and sees the monkey conversing with the lynx.

The miniature is similar in style to fol. 193r. In spite of the effort to achieve a unified composition, the painting is basically divided into two registers, the lower one occupied by the monkey and the lynx, the upper one by the lion who seems to be hurtling down the slope.

Fol. 196r

The monkey advises the suspicious lion to cast off fear and take possession of his territory.

The bright foliage of the trees has a bustling movement typical of the *Ḥamza-nāma* (Pl. 21).

Fol. 197r

The lion, suspecting treachery on the part of the monkey, slays him and flees.

The foliage of the rather large trees is done in clumps, and consists of greenish yellow streaks painted over an indigo ground, the streaks becoming smaller and more indeterminate towards the edges. The technique is common to other miniatures in the manuscript and also to the *Ḥamza-nāma* (Glück, Taf. 43; *Ḥamza,* Pl. 53). The dead and badly mangled monkey lies limply in a corner, while the lion, whose body spans the entire miniature bounds away.

Infra-red photography reveals two layers of pictures beneath the present one (Pls. 115 and 116). The first shows a partially completed picture depicting the parrot addressing Khujasta, similar in style perhaps to a miniature like fol. 102v, and in composition to fol. 179v. We can see clear traces of a building in which is a bed, and provided with kiosks on the terrace, all shown in considerable depth. The parrot's cage is slung from the cornice, and there is the usual enclosed courtyard with an arched door leading to the landscape beyond.

Above this miniature are the beginnings of another painting, the subject matter of which I have been unable to identify. The torso of a standing male figure sketched over a layer of zinc-white is visible to the naked eye where the paint has chipped off over the head of the lion. Infra-red photography also reveals sketches of the heads and torsos of two more figures, not now visible to the eye, on the opposite side of the page about the centre of the left margin (Pl. 116). One of them has an arm thrown out in a gesture of considerable vigour. The other man, right behind this figure, carries something across his shoulders which looks like a wrapped sword of state, or some such object. The drawing of all these figures is quite exquisite, the heads being skilfully modelled.

Above these figures to the left, and also revealed by infra-red photography, are traces of a cursive and only partially legible Persian notation in three diagonal lines, the purport of which is unclear, but which seems to have something to do with the remaking of the picture. According to Professor Naim, the first line contains only one word, *sākht* (he made). In the second line seem to occur the words . . . *tā*? *bīsha* (thicket?) . . .; and in the third line, one word can be read as . . . *siyāh* (black) . . . A diagonal line in ink, apparently drawn by the same pen which wrote the note, is to be seen below what is now the second, and which was probably originally the initial line of the note.

From the above, it is apparent that the first miniature painted on the page was intended to be a parrot and Khujasta scene, invariably found at the beginning of a night's tale, and in the style of the other paintings in this manuscript. The subject, however, was abandoned, as it was in the wrong place, the correct spot for it being on the next folio (198r) where it is indeed to be found. The partially completed picture was therefore covered, and the sketch of another subject drawn, but this too was found to be inconsistent with the text and given up, the present miniature with the correct subject being finally painted over it. The artists must have been quite tired over the trouble this page was causing, and completed it with some haste as can be noticed from the hurried and cursive workmanship.

Cf. fol. 155r which has similar corrections. Several other miniatures also seem to have been corrected, some of them in order to improve the style, and others, perhaps, in order to rectify errors of subject matter. See fols. 36v, 46r, 69v, 73r, and 113r. Infra-red photographs of all these paintings were made, but they reveal nothing more than is visible to the naked eye.

Thirtieth night

Fol. 198r

The parrot addresses Khujasta at the beginning of the thirtieth night.

Though the colour is richer, the painting is derived from the style of the Indo-Persian group with which it shares several features, such as the curving horizon in the corner and the arabesque patterns on walls (Pl. 104). The bright shrubs, similar to those of fols. 193r and 196r, are Mughal interpretations of the plants seen in manuscripts like the Lalbhai *Khamsa* of Niẓāmī (Pl. 102; cf. Pls. 10 and 21). The curious serrated edge of the horizon is also derived from a similar source (Pl. 95). For the ornamental bracket with cage, see remarks under fol. 165v, *supra.*

Fol. 200v

A woman with two children, having abandoned her home, goes into the forest where she encounters a leopard.

The miniature has a marked Mughal flavour in its rhythmical and closely interwoven composition, the movement and gestures of the woman leading to the leopard whose curved body and tail counterbalance the thrust of the trees on top. Cf. fol. 182r to which this miniature has many resemblances. The figure of the lady, with large square head, tasselled jewellery and peg-shaped earring, backless blouse, chequer skirt, and the stiff *oḍhanī,* is an archaistic survival from the *Caurapañcāśikā* group (Pls. 81 and 83), but the harsh angularity has been considerably subdued and the flatness has yielded to a modelled surface.

Fol. 202r

The woman conversing with her children, as the leopard returns, egged on by a fox who is tied to his leg.

Similar to fol. 298r. The chinar tree with its very brightly coloured foliage is seen in the *Ḥamza-nāma,* where it is painted with greater sophistication (Glück, Taf. 10; *Ḥamza,* Pl. 16).

Thirty-first night

Fol. 203v

The parrot addresses Khujasta at the beginning of the thirty-first night.

Attributed to Painter U.

Cf. fol. 209v *et al.* The chequer pattern on the bed is a survival from the *Caurapañcāśikā* group, but the fluttering fringe of the cover is in the early Mughal idiom.

Fol. 205v

The jackal, accidentally dyed an indigo blue, becomes the chief of the animals (Binney collection).

The jackal is enthroned, the weaker animals such as foxes, deer, and wolves being stationed near him, while the powerful animals such as leopards, tigers, lions, and elephants are kept at a safe distance.

This is perhaps the earliest representation of an animal court in Mughal painting, a subject that becomes very popular in the later Akbar period. The composition is fairly simple, the various animals being arranged for the most part in horizontal rows, not yet showing the easy interrelationship and compositional unity characteristic of later miniatures.

Mounds tufted with bright grass fill the ground (cf. Pl. 31), the blades at the edge of the horizon obscuring the sharp transition from it to the sky. The *Ḥamza-nāma,* because of the nature of the story, does not depict many animals, but when it does, the representation is very similar to what we find in the *Ṭūṭī-nāma* (cf. Pls. 31 and 23). Notice particularly the bulging eyes (cf. *Ḥamza-nāma,* Pls. 28 and 31), a convention adopted for all animals. This feature gives way to an increasing naturalism in the post *Ḥamza-nāma* period.

Fol. 207r

The donkey, in a tiger's skin, reveals his identity by braying aloud.

Attributed to Basāvana.

The disguised donkey is at pasture in a veritable sea of billowing grass painted in subtle shades of green and yellow. The furry texture of the tiger skin is caught by skillful use of colour rather than by exquisite attention to detail. The gums and teeth are bared as the donkey brays, instinctively responding to a friend's call. At the edge of the field are two trees and a narrow strip of brown relieved by assured flecks of white meant to represent a fence (cf. fol. 267r), the glowing darkness reminding us of qualities noticed in fol. 35r. Further comparison reveals more similarities including a pronounced preference for rounded forms, a subdued line, free brushwork which is complex but assured, and a soft and wet application of colour, the depiction of grass in this miniature, for example, being exactly paralleled by that of the tree on the lower left in fol. 35. This characteristic, which achieves a true plasticity of colour, is also to be seen in the trees and their trunks in which colour flows as though it were the very sap itself. No Mughal master other than Basāvana worked in quite this manner, and there can be no hesitation in attributing this work to him.

Basāvana has already achieved in this painting, at the very opening stages of his career, all the essentials of his style, which remains basically the same in subsequent years. It is interesting, for example, to compare the tiger in this miniature with one painted by him about thirty years later in the Victoria and Albert *Akbar-nāma* (Welch, "Basawan," Figs. 11 and 13). There, is greater delicacy and refinement, and more restraint, but it is the same beast whose skin the donkey wears in the Cleveland miniature.

Thirty-second night

Fol. 208r

The parrot addresses Khujasta at the beginning of the thirty-second night.

The female figure is reminiscent of Artist FF, cf. fol. 333v.

Fol. 209v

Kaiwān sends a message of love to Khurshid, wife of his brother ʿUṭārid who is away on a journey.

Attributed to Painter U.

The miniature represents a further Mughalisation of the kind of work seen in fols. 67r, 67v, 68r, 68v, and 120r. Several individual details recall the *Ḥamza-nāma,* but because of the miniaturistic scale, the most striking resemblances are to the 1568 *ʿĀshiqa,* particularly in the rich colour, the garden with its streams and cypresses, and the architecture with its domes shaded towards the top. The perspective too is a combination of the frontal and the bird's eye view (cf. Pls. 35 and 36). Vestiges of pre-Akbar styles, however weak, still survive in the bed with its fringe of lappets and tassels (also found in the *Ḥamza-nāma,* cf. Pl. 15), the pattern on the bed cover, and the angular fall of the clothing (cf. Pls. 82 and 84) as seen in the *paṭakās* and scarf ends. These features have been eliminated in the 1568 *ʿĀshiqa,* indicating an earlier stage of development and date for the Cleveland miniature.

Nine other works by this master occur in the manuscript (fols. 203v, 223v, 225r, 227r, 231v, 234r, 241r, 254v, and 335v). Folio 110v represents a similar level of achievement, though it appears to be by another artist.

Fol. 211r

Laṭīf, who has murdered his brother, falsely accuses Khurshid of the deed.

Laṭīf throws up his hands in a gesture of astonishment, the unfortunate Khurshid listening to the accusation with resignation.

Fol. 212v

Unfavourable winds arise on the seas when Khurshid is deceitfully sold into slavery during her pilgrimage to Mecca (Binney collection).

The theme of a storm at sea is popular in Mughal painting, this being its earliest representation. The boat is similar to that found in the *Ḥamza-nāma* (Glück, Taf. 11), as are the expressive human figures. Confusion has broken out on board. A man tugs at the sail with all his strength, another holds a sea-sick comrade, while a third attempts to rescue a man who has already fallen into the water. The serene figure of K̲h̲urs̲h̲id amidst all the tumult emphasises the humility and piety of her character.

The miniature may be contrasted with fol. 125v which is much closer to pre-Akbar painting. Nevertheless, the same lack of connection between the boat and the cabin is to be noticed here, a difficulty that was overcome in the *Ḥamza-nāma* (Glück, Taf. 11; *Ḥamza*, Pl. 17).

Fol. 213v

Kaiwān, Laṭif, and S̲h̲arif, accompanied by ʿUṭārid, at the place of K̲h̲urs̲h̲id who has taken to the life of a devotee.

Attributed to Painter V. Cf. fol. 216v which is identical in style and by the same artist.

The boldly tiled floor is reminiscent of the *Ḥamza-nāma*, as are the human figures (cf. Pls. 24 and 8).

Fol. 215r

The tale of the three men trapped in a cave by a rolling boulder.

In a cave, surrounded by a heaving mass of rocks painted in shades of mauve, violet and brown, relieved with washes of white, are three men, a large boulder blocking the entrance. The rocks, as well as the leafless tree stuck into its crevices, are frequently seen in this manuscript and also in the *Ḥamza-nāma* (Pls. 12 and 20).

The treatment of the hill is derived from the work of K̲h̲wāja ʿAbd al-Ṣamad (cf. Pl. 65) and some of his followers, notably his son Bihzād, examples of whose works are fol. 103r of the *Dārāb-nāma* (Pl. 45) and a miniature reproduced in the Sotheby *Catalogue*, 6 December 67, Lot 111, both of which are ascribed to that painter.

The *Dārāb-nāma* miniature, which has an ascription stating it to be the work of Bihzād, but refined by K̲h̲wāja ʿAbd al-Ṣamad, provides a particularly interesting comparison with this miniature, which may well be by the same painter. The composition is similar, as is the treatment of the rocks, except that these are in much quieter colours and are not as fully modelled, indicating a date in the post-*Ḥamza-nāma* phase. The *Dārāb-nāma* miniature also seems to depict a cave closed by a boulder, the mouth of the cave actually sealed by a rock, while here a portion of the rock has been omitted in order to allow us to see the figures inside. This concern for maximum visibility is a characteristic of both the *Ḥamza-nāma*

(Pl. 18) and this manuscript (fols. 29r, 47v, and 49v), where incidents taking place in a well or in a dungeon are revealed by a similar device.

Fol. 216v

Khurshid reunited with her husband ʿUṭārid.

Attributed to Painter V. Cf. fol. 213v which is by the same artist.

Thirty-third night

Fol. 217r

The parrot addresses Khujasta at the beginning of the thirty-third night.

Attributed to Painter W.

The style of the painter is similar to that of Painter U (cf. fol. 223v), but is not as refined in the execution of detail, and archaistic features survive to a greater extent. The female type is distinctive in dress and jewellery, a peg-shaped earring being preferred.

Fol. 219v

Preparation for the marriage of Maḥmūda, daughter of the grand vizier, and Ayāz, son of the second vizier.

Attributed to Painter Z.

The miniature has many resemblances to fol. 225r attributed to Painter U (trees, architecture, seated figures, particularly the bearded man), but the workmanship is not as fine, and the composition of the figures is more static and mechanical.

The heavy red curtain can be compared to the green curtain of fol. 124r, which it resembles, and to Basāvana's more assured handling in fol. 33v. Curtains of this type occur in the *Ḥamza-nāma* (Glück, Taf. 30, Abb. 20; *Ḥamza,* Pl. 12) and the 1570 *Anwār-i Suhailī* (Pl. 40), where the shading is more accomplished. For similar types in the *Ḥamza-nāma* see Pl. 8.

Fol. 221r

Hearing her declaration of love, Ayāz falls at the feet of Maḥmūda at the holy shrine. The scene is witnessed by Salīm, Ayāz's friend, and a maid.

Attributed to Painter W.

Fol. 222r

Salīm and Salīma return to Ayāz and Maḥmūda in the sanctuary.

Attributed to Painter X.

The style of the painter is very close to that of Painter U, more than to Painter Z, but the scale of the figures is larger, the detail less exquisite. These features, however, might well be explained as the consequence of hastier workmanship, so that it is conceivable that Painters U and X are one and the same person.

Cf. fol. 335v by Painter U, where both the similarities and differences between the two painters are particularly clear.

Fol. 222v

The two couples reach a foreign city where they make their home.

Attributed to Painter X.

Identical in style to fol. 222r.

The city, with its many kinds of buildings, tiled walls, and moat traversed by a diagonally placed bridge, is like those in the *Ḥamza-nāma* (Glück, Taf. 5; *Ḥamza,* Pl. 7). The composition with a projecting and prominent city gate, the grassy terrain, and the rocks at the lower margin, is similar to the *Ḥamza-nāma* (Glück, Taf. 41; *Ḥamza,* Pl. 48). The only difference is in the more minute and ornamental execution of detail in the *Ḥamza-nāma,* made possible by its much larger size and the ambitious and exceptional nature of the undertaking, and is to be understood as one of quality rather than kind. Pre-Akbar traditions are in little evidence, very faint survivals to be noticed in the striped dome above the city gate, traceable to the *Candāyana* group both here and in the *Ḥamza-nāma* (cf. Pl. 17) and the mauve and maroon chequer pattern on the lady's skirt, which is of *Caurapañcāśikā* derivation.

Thirty-fourth night

Fol. 223v

The parrot addresses K͟hujasta at the beginning of the thirty-fourth night.

Attributed to Painter U.

Both the figure of K͟hujasta and the parrot are exquisitely finished. The details are accurately observed and minutely rendered, including a finely drawn cord that is secured around the bird's leg.

The painter works painstakingly on human and animal figures, while the architectural decor or the landscape, though rich in colour, is somewhat freely painted. This serves to fix our attention on the main figures, the delicacy of their execution being further enhanced by the contrast. Cf. Pl. 5, a detail from the *Ḥamza-nāma.*

Fol. 225

The three young men present themselves as suitors for the hand of Zuhra, the daughter of a merchant of Kabul.

Attributed to Painter U.

The expressive faces, delicately modelled near the eyes, nose, and mouth, with clearly drawn and slightly pouting lips, are typical of the painter. In spite of the refined line, freedom and fluency is not lost, observable even in small details like the drawing of the turbans of the suitors. Emotional states are also explored. The suitors are seated in postures of great ease, their self-assured bearing and looks contrasting with the worried face of the father.

Painter U lacks the keen sensitivity to plastic form displayed by Basāvana, though the robes of the father, and the accomplished brushwork of the sky and tree suggest some affinity of style. Cf. fol. 267r. For similar, but more crudely drawn faces in the *Ḥamza-nāma* cf. Pl. 8.

Fol. 226r

The third suitor, who is an archer, shoots the wicked fairy who has imprisoned Zuhra. He rides on a magic horse prepared by the second suitor and is led to the spot by the divining prowess of the first.

The composition is strongly reminiscent of the *Ḥamza-nāma* (Glück, Abb. 3), with the same flair for dramatic combat, and a very similar treatment of rocks (Pl. 12). The fairy, wearing a long tunic and a coat with wings sprouting from the shoulders, is derived from Safawi painting (Binyon, *Nizami,* Pl. XIV) and almost identical with those occurring in the *Ḥamza-nāma* (Pl. 13) and the 1568 *ᶜĀs̲hiqa* (Pl. 36).

Fol. 227r

The Rājā's son vows to sever his head, and offer it to the image if he is united with the princess he has seen in the temple.

Attributed to Painter U.

Cf. fol. 209v, the figures of the seated ladies being identical. The softly modelled faces are exquisitely painted.

The image is painted all in gold, and is related to those seen in fol. 20r. They have, however, narrow waists and exaggerated chests recalling Western Indian prototypes, while these resemble more the *Ḥamza-nāma* examples (Pl. 11). The temple also is not a symbolic representation, but a domed building with a pot-shaped finial and a flag.

Fol. 229v

The princess discovers the dead bodies, with heads severed, of her husband and his Brahman friend.

Attributed to Painter X.

Cf. fol. 222v. Also notice the cursive painting of the background and the foliage of the trees. The waving sky outlines the dark horizon sharply, a feature shared with fol. 227, attributed to Painter U. There, however, the contrast is not so harsh and is further softened by the stippled trees.

A flag was originally planned for the temple, but omitted, the drawing being still visible.

Thirty-fifth night

Fol. 231v

The Brahman gives an account of his falling in love with the king of Babylon's daughter to his friend, the magician.

Attributed to Painter U.

The artist has painted both the occasion when the Brahman falls in love, smitten by the woman's glance in a garden, and the Brahman's subsequent recounting of this incident to his friend, in the same miniature.

Also cf. fol. 209v by the same painter. He shows great skill in capturing the light and colour of gardens, though he does not lavish the same careful detail on them as he does on human figures.

Fol. 232v

The magician, disguised as a Brahman, visits the king of Babylon.

He is accompanied by his Brahman friend who has changed into a woman by virtue of the magic bead which he holds in his mouth. The magician passes her off as his daughter-in-law.

The painting reminds one of work by Painter X in the modelling of the heads (fol. 229v), but the work is coarser, the heavy-jawed figures being more in the style of fol. 269v.

Fol. 234r

The son of the king of Babylon sees the Brahman transformed into a woman bathing and falls in love with her.

Attributed to Painter U.

The expressive figure of the princess, who leans forward suddenly, is balanced by the arm that points to the figure on the terrace. Also cf. fol. 225r, where sky, trees, and grassy ground are identically painted.

Fol. 235r

Consternation in the palace of the king of Babylon at the disappearance of the princess and her companion (Private collection).

Attributed to Painter Z.

The miniature is similar to fol. 219v, and is clearly distinguishable from the work of Painter U by virtue of its coarser workmanship. Painter Z is a follower of Painter U, but his work is much less accomplished.

Fol. 236r

The magician disguised as the Brahman returns to claim his "daughter-in-law."

On not finding her, he threatens to stab himself and is thereupon offered an indemnity in cash which is being carried in by an attendant. The sacred thread lies torn on the ground. The heavily tiled floor, serving as a backdrop for the action, is similar to those found in the *Ḥamza-nāma* (Pl. 24).

Thirty-sixth night

Fol. 237r

The parrot addresses K͟hujasta at the beginning of the thirty-sixth night.

The rough execution of this miniature is evident when compared to fol. 223v, which is much more delicate and refined in its execution.

Fol. 238r

The four viziers at the house of a merchant of Zavol to see his daughter Maḥrūsa who has been proposed in marriage to the king (Binney collection).

Attributed to Painter Z.

Cf. fol. 235r.

Fol. 239v

The king dreams of a lady, the personification of wealth, departing from him on account of his purchasing a bowl and a staff from a yogi.

Probably by Painter Y.

The gauzy texture of the diaphanous fabric worn by the lady can be contrasted with fol. 332v attributed to Painter FF. The work is almost indistinguishable from that of the *Ḥamza-nāma.*

Fol. 240v

Maḥrūsa's marriage to the prefect of the city.

Attributed to Painter Z.

Cf. fol. 219v. The figures are more spaced out, the composition flatter, and the workmanship coarser. This is particularly evident in the drawing of Maḥrūsa and her old duenna, compared to Maḥmūda and her duenna in fol. 219v.

Similar figural and facial types occur in the *Ḥamza-nāma*, (Pl. 8).

Fol. 241r

The king of Babylon sees Maḥrūsa from his palace balcony.

Attributed to Painter U.

Cf. fol. 231v, the male and female figures being identical. The architecture, the prominent tiling, and the tree with stippled foliage in the background are as in the *Ḥamza-nāma* (Pls. 15, 24 and 20).

Fol. 242r

Maḥrūsa kills herself at the tomb of the king of Babylon, and her husband does likewise.

Similar in style to fol. 221, which is attributed to Painter W, but the drawing here is more delicate.

Thirty-seventh night

Fol. 243r

The parrot addresses Khujasta at the beginning of the thirty-seventh night.

Attributed to Painter W.

Cf. fol. 217v, which is identical in style.

Fol. 245v

The prince, a son of the ruler of Sistan, enters the service of a snake.

Fol. 247r

The prince enthroned as the ruler of Sistan (Keir collection).

Attributed to Painter Z.

Cf. fol. 235r.

Thirty-eighth night

Fol. 248r

The parrot addresses Khujasta at the beginning of the thirty-eighth night.

By the same artist as fol. 239v, probably Painter Y. Note the similar treatment of the soft cloth worn by the female figures.

Thirty-ninth night

Fol. 254v

The queen of Rūm watches the peahen prefer to burn rather than abandon her eggs while the peacock flees the nest.

Attributed to Painter U.

Cf. fols. 231r and 241r. The burning tree is painted with great subtlety and skill. It is the foliage and upper branches that have caught fire, the flames licking the smoking nest, as burning leaves fall to the ground. The inevitable destruction of the tree is indicated by the deliberate smearing and erasure of the pattern of its foliage.

Fol. 256v

The painting made by the vizier of the emperor of China for the queen of Rūm.

The miniature reproduces the painting done by the vizier. It depicts a doe abandoning the stag and its young ones in a flood, while the emperor of China looks on.

Stylistically the miniature represents a phase where the artist is beginning to closely approach the *Ḥamza-nāma* style, but has not yet been able to overcome the awkward vestiges of earlier tradition. The lavender-coloured hill with tufts of sparse grass is of Persian or Indo-Persian derivation (cf. Lalbhai, *Kh̲amsa* of Niẓāmī, Pl. 99), but the trees (cf. Pl. 20) and swirling water are in the *Ḥamza-nāma* manner. The design of the building with its projecting balcony approaches a type seen in the 1568 *ʿĀshiqa* (Pl. 35), though lacking its elegance.

Fortieth night

Fol. 258r

The parrot addresses K̲h̲ujasta at the beginning of the fortieth night.

Attributed to Painter AA.

This painter made several miniatures towards the end of the manuscript. The style is like that of Painter Y, but somewhat more crude, emphasising large heads and exaggeratedly heavy jaws which are of pre-Akbar derivation (cf. Pls. 67 and 68) and which also occur in several of the *Ḥamza-nāma* paintings (Pls. 24 and 27). The lady in this miniature, with characteristic curving line of the jaw, is a good example of his manner of painting female figures.

Fol. 259v

Sha̲hr-Ārāī and her lover dallying on a bed beneath which is concealed her husband.

   Attributed to Painter W.

Fol. 261v

Sha̲hr-Ārāī's husband bends to kiss his wife who feigns sleep.

   Attributed to Painter AA.

   The head of the bearded husband, large in proportion to the rest of the body, and extremely heavy-jawed, is characteristic of Painter AA. It also occurs frequently in the *Ḥamza-nāma* (Pls. 27 and 28).

Fol. 262r

Sha̲hr-Ārāī and her husband adopt her lover as a brother in the family.

   Attributed to Painter Y.

   The artist, both in the large proportions of his figures and the bold and vigorous depiction, is working in a style indistinguishable from the *Ḥamza-nāma*, hardly any traces of non-Mughal styles being evident in his work.

Forty-first night

Fol. 263r

The parrot addresses K̲hujasta at the beginning of the forty-first night.

   Attributed to Painter AA.

   Cf. fol. 258r with which this miniature is identical in style. In spite of the proximity to the early Mughal style, the figure of the lady being close to representations in the *Ḥamza-nāma* (Pl. 2), the stiffly tasselled scarf edge of the *Caurapañcāśikā* group survives here as in fol. 258r, though this feature has all but disappeared from the *Ḥamza-nāma* (Pl. 4).

Fol. 267r

The gardener seizes and beats a donkey who insisted on braying, while the deer, its companion, flees to safety.

   Attributed to Painter X.

   The miniature can be compared with fol. 207r, attributed to Basāvana, with which it shares several common features, notably the same types of trees and grassy meadow. Even the strip of brown fence, here separating the pink earth from the green grass, is present. Nevertheless, the ease and verdant bloom of Basāvana is not attained, nor his plasticity of colour.

The Beatty *Ṭūṭī-nāma* version of this incident (Pl. 58) is an adaptation of this composition, the deer placed in the background, the donkey thrashed by the gardener in the foreground. The luxuriantly brushed grass is replaced by stippled green dots and the fence is represented with clearly painted rows of embedded branches. The softly painted foliage of the Cleveland miniature is also abandoned, though its memories survive in the tree to which the donkey has been secured.

Forty-second night

Fol. 268r

The parrot addresses Khujasta at the beginning of the forty-second night.

Attributed to Painter Y.

Fol. 268v

The marriage of ʿUbaid, son of a merchant of Tirmiz.

Attributed to Painter Y.

Cf. fol. 262r, particularly the seated female figures.

Fol. 269v

The merchant of Tirmiz takes the wise parrot and myna to ʿUbaid.

Attributed to Painter AA.

The heavy heads have particularly close parallels in the *Ḥamza-nāma* (Pl. 28).

Fol. 271r

The Rājā's daughter, born with three breasts, accompanies her blind husband and his hunchback guide on a journey.

Attributed to Painter Y.

The group of three figures, consisting of the blind man (cf. *Ḥamza-nāma*, Pl. 6), holding on to his wife with one hand and the staff which is held by the hunchback in the other, is masterfully arranged, sensitively conveying their slow but steady progress on the journey. The portraiture is also very expressive; the bewildered look of the unfortunate princess, the helpless face of the blind man, and the hopefully expectant face of the hunchback, who points ahead as though he has sighted their goal in the distance, are among the finest in the manuscript.

Fol. 272v

The Rājā's daughter and her lover stoned to death for adultery.

Gory violence, a characteristic of the *Ḥamza-nāma* style, is only too evident here.

The raised hands, all holding stones and surrounding the couple, emphasise the relentless brutality of the execution. The badly mutilated figure of the woman lies crumpled and dead, all traces of life having vanished from her face and body. The man, hair dishevelled and pain on his face, awaits death calmly.

Fol. 275r
Repenting his conduct, ʿUbaid falls at the feet of his parents.
Attributed to Painter X.
Cf. fol. 229r, also attributed to this painter.

Forty-third night

Fol. 276r
The parrot addresses Khujasta at the beginning of the forty-third night.
The miniature, particularly the figure of the woman, is reminiscent to some extent of the work of Painter X. Cf. folio 229v.
A comparison with a painting of the same subject in the Beatty *Ṭūṭī-nāma* (Pl. 59) shows similarities of composition with a door on the side and back walls, the lady being placed in the center. The Cleveland miniature, however, is indistinguishable from the style of the *Ḥamza-nāma* in its vivid colour and ornamental decoration of architecture, and much earlier than the Beatty example where both the colouring and the decoration are considerably restrained.

Fol. 278v
The snake, hidden in a basket of flowers, reveals himself to the Rājā who has just sent away his wife.
The recumbent Rājā, covered by a diaphanous bedsheet, props his head as he converses with the snake. Problems of depth are handled with ease in the depiction of the Rājā and the bed on which he reclines. The flat treatment of the architecture, particularly the upper portion, is at odds with this conception of space. The triangular blue clouds bordered by wavy white lines and the domed kiosk are unusually archaistic, reminiscent of the *Caurapañcāśikā* group (Pl. 76) and even earlier traditions (Pl. 68). So is the lady with curvaceous body and the sharply flaring *paṭakā,* who can be compared with similar types in the *Ḥamza-nāma* (Pl. 2).
A miniature in the National Museum, New Delhi (Pl. 60), a stray leaf from the Beatty *Ṭūṭī-nāma,* represents the same scene, and is clearly based on the Cleveland miniature. The posture of the Rājā, leaning against the bolster with legs crossed and head propped, is repeated. Significantly enough, this posture does not appear

elsewhere in the Beatty manuscript, the parallel further confirming its close dependence upon the Cleveland manuscript. The upper part of the building has also been given depth.

Forty-fourth night

Fol. 282v
The parrot addresses Khujasta at the beginning of the forty-fourth night.

Attributed to Painter BB. Cf. fol. 287v which is by the same artist.

The figure of the lady, with large eyes, squarish face, stiffly projecting scarf, and large flowers in her braided hair, is closely comparable to miniatures of the *Caurapañcāśikā* group (Pl. 84), and forms the basis of a more accomplished version in the *Ḥamza-nāma* (Pl. 4). Other similarities are the pavilion with canopy secured by tassels, the bed with bolster placed on edge, and the vessels below. Even the decorative flowers placed above the bolster are repeated here, though they rest more plausibly on the bolster itself (also cf. fol. 102r). The knotted bracket and the *makara* gargoyle with a three-pointed flag decorated with tassels are also present (Pls. 82 and 84).

Inspite of these similarities, the bright colour is conspicuously absent, the pale tonalities of the *Candāyana* group being preferred.

Cf. fol. 43r.

Forty-fifth night

Fol. 286r
The parrot addresses Khujasta at the beginning of the forty-fifth night (Binney collection).

Fol. 287v
The Amīr slays the snake after giving it shelter.

Attributed to Painter BB.

Two incidents are depicted. In the background is the owner asking the Amīr the whereabouts of the snake, the snake itself partially visible under the robe. In the foreground, the Amīr kills the snake by dashing it to the ground.

The miniature is by the same artist as fol. 282v, the facial features being identical. It provides an interesting illustration of the new types of composition and work being attempted by an artist entering the Mughal atelier from the *Caurapañcāśikā* tradition.

Forty-sixth night

Fol. 291v

The court of the Rājā of Ujjain.

Attributed to Painter Z. Cf. fols. 240v, 247r. Notice the cursive colouring of the cypresses and mounds of grass.

According to the text, the Rājā had just hunted a strange animal, the softness of whose body was being favourably compared with the fur of the sable and the ermine. The composite animal with the beak of a bird, ram's horns, and a furry skin is the artist's version of the fabulous beast.

The miniature is close to the *Ḥamza-nāma,* particularly in the vigorously gesticulating figures and grimacing faces, especially those with distorted mouths and bared teeth (cf. Pls. 8, 27, and 30). The central figure, right row, in partial two-third profile with sharp nose and pointed chin is a direct adaptation from the Western Indian style or its variants, and the same type is also retained in the *Ḥamza-nāma* (Pl. 22).

Fol. 292v

The parrot laughs on hearing the Rājā of Ujjain's wife admire her beauty in a mirror.

Attributed to Painter CC. Cf. fol. 315r which is by the same artist.

The bright colour and the depth of composition are Mughal in character, but the figure of the lady is strongly reminiscent of *Caurapañcāśikā* traditions as seen in other miniatures of the manuscript.

Fol. 293v

The Rājā of Ujjain, who is travelling in the guise of a yogi, meets two brothers who ask him to equitably partition their father's possessions.

Attributed to Painter DD. Cf. fols. 308v and 325v which are by the same artist.

The water is painted in unusually stiff swirls (cf. fol. 306r), and is edged by fairly evenly spaced boulders provided with circlets of leaves. The rocks are very heavily outlined.

Forty-seventh night

Fol. 300v

The parrot addresses Khujasta at the beginning of the forty-seventh night.

The miniature is heavily dependent on *Caurapañcāśikā* traditions, retaining the monochrome red and green patches of colour (Pl. 81). Cf. fol. 43r. The figure of the lady recalls that on fol. 328r.

Fol. 301v

The four destitute friends go to a wise man who gives each one of them a magic shell to be placed on top of the turban.

Cf. fol. 72r which this miniature resembles, though it represents a closer approach to the Mughal manner.

Fol. 302v

The fourth man digs at the spot where he dropped the shell, expecting jewels, but discovering mere iron.

The ground is littered with boulders. In the hills are deer, a mountain goat, and, to the extreme left, a bear holding a large rock. This motif, which has nothing to do with the story, is a quotation from Safawi painting. Cf. *Bahrām hunting the lion* ascribed to Sulṭān Muḥammad in the British Museum *Kh̲amsa* of Niẓāmī dated 1539–43 (Binyon, *Nizami,* Pl. XV). Mīr Sayyid ʿAlī, in whose atelier the Cleveland *Ṭūṭī-nāma* was prepared, also, significantly enough, worked on that manuscript.

Forty-eighth night

Fol. 305r

The young man of Baghdad solicits advice from a friend at the behest of his slave-girl who is adept at music.

The artist attempts to lend depth to the pavilion, but not very successfully, the four pillars being on the same plane.

Fol. 306r

The bag of gold which he received for the slave-girl being stolen in a mosque, the young man of Baghdad tears his clothes and is about to fling himself into the Tigris.

Attributed to Painter EE. Cf. fol. 307r which is by the same artist.

The city with its mosque and buildings is reminiscent of the *Ḥamza-nāma* (Pl. 33), but the workmanship is quite unrefined. For a similarly composed *Ḥamza-nāma* painting, see Grube, *Islamic Paintings,* Col. Pl. XLIX.

Fol. 307r

The young man of Baghdad joins the Hās̲h̲imī's boat as a sailor to find his slave-girl on board.

Attributed to Painter EE. Cf. fol. 306r which is by the same artist, the treatment of the hills, water, and the figures being very similar.

Cf. Beatty *Ṭūṭī-nāma* (Pl. 61) which shows the same scene. The cabin is more securely placed in the center of the boat, the fewer figures are more comfortably arranged and more delicately rendered. The treatment of the background with the hills is also less like the *Ḥamza-nāma,* the trees painted in green and yellow, the bright rocks bordering the water in the Cleveland miniature being omitted.

Fol. 308v

The young man of Baghdad reveals his true identity to the Hāshimī.

Attributed to Painter DD.

The same stiff treatment of water, inhabited by large fishes and bordered by rocks circled with leaves, identifies this miniature as the work of the artist of fol. 293v.

Fol. 310r

The young man of Baghdad reunited with his slave-girl.

The space left for the sky has not been painted. Note the dark and heavy use of colour.

Forty-ninth-night

Fol. 311r

The parrot addresses Khujasta at the beginning of the forty-ninth night.

The workmanship is reminiscent of fol. 324v, and the miniature could perhaps be by Painter FF.

Fol. 315r

The eldest brother explains the reasons for his youthful appearance.

Attributed to Painter CC. Cf. fol. 292v which is by the same artist, the figure of the woman being identical. This type of female figure, with distinctive blouse pattern and hair braided with pearls, is very similar to those in the *Ḥamza-nāma* (Pl. 2).

Fiftieth night

Fol. 316r

The parrot addresses Khujasta at the beginning of the fiftieth night.

Attributed to Painter FF.

The artist prefers the use of tiled floors as a backdrop for his figures, in this and in other respects resembling the *Ḥamza-nāma* (cf. Pl. 24). Female faces are quite distinctive, with small nose and slightly bulging forehead, on which fall loose strands of hair, and modelled accents at nose and chin. A sharply projecting *paṭakā* is the only vestige of earlier traditions.

Fol. 317r

The king's emissary being provided with gifts for his mission to Rūm in order to seek the hand of the emperor's daughter in marriage.

The style is similar to fol. 301r, not very accomplished, but beginning to show early Mughal features, such as the backdrop of a tiled floor for the figures working in the foreground. The movement is unsure, the flowing rhythms that unite one figure to another being not yet achieved.

Fol. 318v

The merchant returns bringing a young slave who is really the son of the princess of Rūm, now married to the king.

The elaborate architecture recalls that of fol. 256v, and the miniature is probably by the same artist.

Fol. 321v

The guard spares the life of the slave when he learns that he is the son of the princess of Rūm.

Attributed to Painter AA.

Fol. 323r

The king places the talisman on his sleeping wife.

Attributed to Painter AA.

The reclining lady, with large and heavy head, can be compared to fol. 261v, also by the same painter. The bearded king is very close in style to the guard flourishing a sword in fol. 321v.

Fol. 323v

The guard restores the son who falls at his mother's feet.

Attributed to Painter AA.

An ascription on the lower margin has been cut off.

Though painted with less attention to detail, the garden with a row of alternating bananas and cypresses immediately recalls the *Ḥamza-nāma* (Pl. 32). Softness and texture is imparted to the lady's clothing, but the pattern on the skirt and the tasselled edge of the *oḍhanī* are faintly reminiscent of *Caurapañcāśikā* traditions, as are other works by this artist (cf. fols. 258r and 263r).

Fifty-first night

Fol. 324v
The parrot addresses K̲h̲ujasta at the beginning of the fifty-first night.
    Attributed to Painter FF.
    The figure is smaller in scale than fol. 316r, but is clearly by the same artist.

Fol. 325v
K̲h̲ulāṣa, a vizier, sees the daughter of K̲h̲aṣṣa, another vizier, and covets her.
    Attributed to Painter DD. Cf. fol. 308r, the treatment of the human figures and the trees being very similar.

Fol. 328r
King Bahrām, who has married K̲h̲aṣṣa's daughter, has her tied to a camel to be abandoned in the desert as a result of false accusations made by K̲h̲ulāṣa.
    Attributed to Painter GG. Cf. fols. 339r and 340r which are by the same artist.
    The tiled floor provides the backdrop for the scene. The workmanship is fairly coarse. The king is seated with one knee raised on a throne, the legs of which are curiously bent and projecting.

Fol. 329v
K̲h̲usrau, the King of Kings, pays homage to the pious daughter of K̲h̲aṣṣa.
    Attributed to Painter W.
    The dark shading of mounds tends to break up the field into horizontal strips. The well is not sunk into the ground, but stands on it, a section of the brick lining not being painted, in order to depict the cavity within. Cf. fol. 47v.

Fifty-second night

Fol. 332v
The parrot addresses K̲h̲ujasta at the beginning of the fifty-second night.
    Attributed to Painter FF.
    Cf. fols. 316r and 324v. Note the mauve-and-maroon pattern of the skirt. This

type of figure, the angularities further toned down, occurs in the *Ḥamza-nāma* (Pl. 2).

Fol. 333v

The bird of seven colours brings a sable to the pious man.

Attributed to Painter FF.

The boldness of the tile pattern is subdued by its pale colour, salmon pink and gold. A very similar bird occurs in the *Ḥamza-nāma* (Glück, Abb. 36; *Ḥamza,* Pl. 52).

Fol. 335v

The pious man's wife offers the seven-coloured bird as food to her lover, but not finding its head, he breaks the pot and bowl in anger.

Attributed to Painter U.

Cf. fols. 241r and 254v.

Fol. 337v

The son of the pious man slays the dragon.

The pale ground and hill, painted in the same colour, are rather unusual.

Fol. 338r

The king asks the pious man's son for the whereabouts of the dragon.

The colour is similar to fol. 337v and suggests that it may be by the same artist.

Fol. 339r

The king gives his daughter in marriage to the pious man's son.

Attributed to Painter GG. Cf. fol. 328r. The posture of the seated king and the attendant waving a scarf immediately behind him are identical in the two miniatures.

The king converses with the pious man's son behind whom is seated his daughter. Women are bringing in gifts on trays, while two female musicians play on the lute and the tambourine. The arrangement, though not lacking in depth, is fairly stilted, the whole being painted against a patch of flat, unrelieved mauve. Traces of archaism are also noticeable in the stiff fall of the women's *oḍhanīs* and the sharply projecting scarves.

The mauve ground and the seated musicians are Indo-Persian features derived from miniatures like those in the Lalbhai *K͟hamsa* (Pl. 101).

Fol. 340r

The pious man's son, now a king, reveals himself to his father. His nurse upbraids his unfaithful mother.

Attributed to Painter GG. Cf. fols. 328r and 339r. The figures, particularly the standing mother wearing a scarf with stiff tassels, and also the throne with projecting legs are similar in the two miniatures.

CHAPTER V
CONCLUSION: THE PLACE OF THE ṬŪṬĪ-NĀMA IN EARLY AKBAR PERIOD PAINTING

In the earlier chapters, we have attempted to provide the background for a proper evaluation of the Cleveland *Ṭūṭī-nāma* and its place in the history of sixteenth century painting by a brief discussion of the traditions of painting and patronage at the courts of Bābar and Humāyūn, of what has been called pre-Akbar Indian painting, and also of the Mughal style proper of the early Akbar period. We followed this with a somewhat detailed description and stylistic analysis of the numerous miniatures of the *Ṭūṭī-nāma,* taking care to draw attention to its relationship with the various types of pre-Akbar Indian painting on the one hand, and early Akbar painting, especially the *Ḥamza-nāma,* on the other. While we are now thus in a position to draw the necessary conclusions to our study, before we do so, it is proper to examine views opposing our own, expressed since their preliminary publication some time ago. It would also seem important to do this in order to prepare the ground for a clear and unimpeded presentation of our own views as they stand now.

Responses to the preliminary report

The *Ṭūṭī-nāma* was first published, shortly after its acquisition by the Cleveland Museum of Art, in a preliminary note which Sherman E. Lee and I contributed to the Burlington Magazine.[1] There, we tentatively proposed that the manuscript was "not of a later period than the *Ḥamza-nāma,*" and that it most likely represented "in large part the earliest, most archaic and formative phase of Akbar's atelier."[2] We were led to this conclusion by the rather striking presence of pre-Akbar traditions alongside elements of the *Ḥamza-nāma* style, a peculiarity which led us also to suggest that the *Ṭūṭī-nāma* could have been begun "slightly before the *Ḥamza-nāma* was commissioned" and that it was completed while that "large scale monument of manuscript illumination was well under way."[3] The dates of the *Ḥamza-nāma* being uncertain, we suggested a range of 1560–68 for the *Ṭūṭī-nāma* primarily with reference to the 1568 *ʿĀs͟hiqa* which it would appear to pre-date. We also noticed that though several paintings were not of a sufficiently high quality, the manuscript

1. Lee and P. Chandra, pp. 547–54.
2. *Ibid.,* p. 553.
3. *Ibid.,* p. 550.

itself was a product of the early Mughal studio, as other Mughal manuscripts possessed a comparable range of work. The presence in the *Ṭūṭī-nāma* of authentic works of well-known masters of the school, such as Basāvana and Dasavanta, further supported this conclusion.

The detailed study presented in this book largely confirms and strengthens our original hypothesis, though it has also been necessary on occasion to adjust our previous views. The most important reconsideration is, perhaps, the thought discussed below, that though certain parts of the Cleveland *Ṭūṭī-nāma* may be stylistically anterior to the *Ḥamza-nāma*, this may possibly not be the case in terms of the actual chronology.

Stuart Cary Welch noticed the Cleveland *Ṭūṭī-nāma* shortly after its initial publication in his catalogue of an exhibition of Mughal art, which included some *Ṭūṭī-nāma* miniatures, held in New York in 1964,[4] and he basically agreed with our views. This was followed by an intemperate attack upon him by Karl Khandalavala in an article that is unvarnished polemic rather than a work of reasoned scholarship.[5] Characterising Welch, for example, as a "comparative newcomer,"[6] he declared at least two drawings and one painting of the Akbar period published by Welch to be modern copies. The only reasons Khandalavala advanced for this opinion was a reference to his "constant contact with the Indian markets over a long period of years," and information provided to him by a Delhi dealer some time in 1925–30, according to whom the pictures in question were copies, and so be it. These somewhat unorthodox criteria were also extended to the *Ṭūṭī-nāma,* which savoured to him of the methods of the "copyist Appaya" of Hyderabad, who, according to Khandalavala, "specialised in producing and restoring illustrated manuscript folios in the early Mughal style," also forging on to them ascriptions to painters whose names were "culled from the *Āʾīn-i Akbarī.*"

Basil Gray, in a review of the Welch catalogue,[7] though admitting the presence of "Sultanate" features, suspected the *Ṭūṭī-nāma* to be not earlier than 1580 on the basis of what he calls European influence, such as the shading of curtains and garments, as seen in fol. 33v. Even if we were to grant for the sake of argument that these features are of European origin (and this has yet to be established), there are no solid reasons to assume that Mughal painters had not become acquainted with European painting earlier. The first recorded instance of Akbar coming into contact with the Portuguese, and there may have been earlier unrecorded ones, is during the

4. Welch, *Art of Mughal India,* pp. 24–25, 162.
5. Khandalavala, "Some Problems of Mughal Painting," pp. 9–13. Though the article is carried in a *Lalit Kala* issue dated 1962, it was obviously written in 1964–65.
6. *Ibid.,* p. 12.
7. *Artibus Asiae* 28, p. 100.

Gujarat campaign of 1573, a full seven years before the Jesuit mission of 1580; and we have an account of the return of what seems to have been primarily an artistic mission to Goa under Ḥājī Ḥabībullāh in 1575. In the painting itself, and contrary to Gray's assertion,[8] shaded curtains and garments are seen in the 1570 *Anwar-i Suhailī* (Pl. 40) and also in the *Ḥamza-nāma* (Pl. 25), where in one painting also occur figures wearing European dress (Pl. 14). As both these manuscripts fall in the same broad range of time as the *Ṭūṭī-nāma*, the "European influence" argument for a 1580 dating of the manuscript will not hold.

The date of ca. 1580 for the *Ṭūṭī-nāma* was also suggested to Gray by the presence of a miniature by Basāvana (fol. 36v), who, according to him, was an artist "of the late Akbar period," his first works being represented in the Jaipur *Razm-nāma,* dated for unstated reasons to 1584–9,[9] and in the British Museum *Dārāb-nāma* dated to the 1580's. Gray also felt that, as the *Ṭūṭī-nāma* miniature did not appear to differ in date from the *Dārāb-nāma* example, the manuscript had to be put "firmly in the 1580's." Now there is no evidence that would support Gray's assertion that Basāvana was a late artist. If the career of another painter, namely Khwāja ʿAbd al-Ṣamad, extended from at least as early a period as the time of his joining Humāyūn at Kabul in 1549 to as late as 1595, the date of his last known painting in the British Museum *Khamsa* of Niẓāmī, a minimum period of forty-six years, there is nothing inherently impossible in Basāvana's working in the atelier from 1560 to the closing years of the reign.[10] Also Gray's opinion that there was no difference in the date of Basāvana's *Ṭūṭī-nāma* and *Dārāb-nāma* miniatures is not backed up by a consideration of their style.[11]

Khandalavala, presumably with the support of Jagdish Mittal, reverted to the subject of the Cleveland *Ṭūṭī-nāma* once again in connection with the miniature ascribed to Dasavanta (fol. 32v), the argument running as follows:[12] According to the *Āʾīn-i Akbarī,* Dasavanta was discovered by Akbar and handed over for training to Khwāja ʿAbd al-Ṣamad; and as the royal protégé could be trained by none other than the head of the imperial atelier, a contrary situation not being consonant with "Mughal royal dignity," Dasavanta's training could only have begun after the Khwāja became the head of the atelier, and this (according to Khandalavala's dubious analysis of the *Nafāʾis al-Maʾāsir,* as demonstrated by us earlier) was only in

8. *Ibid.:* "There is not a trace of it [European influence] in the well-known *Anwar-i Suhayli* of 1570."
9. See *supra,* p. 68, fn. 51.
10. The last known work by Basāvana would probably be the signed border in the *Muraqqaʿ-i Gulshan* referred to by Wilkinson, "Indian Paintings in a Persian Museum," p. 173. The example is not dated but the compilation of the *Muraqqaʿ* probably did not begin much before Ramaẓān 1008/ March 1600 which is the earliest dated margin to be discovered. See Godard, p. 13.
11. See *supra,* p. 93.
12. Khandalavala and Mittal, "MS. of Tilasm and Zodiac," p.16.

1574.[13] Dasavanta could not thus have entered the ranks till at least 1576. Therefore, all paintings by him would have to be after 1576; and as the *Ṭūṭī-nāma* contains a painting ascribed to Dasavanta, it must be later than 1576.

The argument makes no sense. Aside from the uncertainty regarding the exact date when Khwāja ʿAbd al-Ṣamad took charge of the *Ḥamza-nāma* (if this in fact meant becoming the head of the atelier),[14] asserting the impossibility of the royal protégé being trained by the Khwāja before he became the head of the atelier is not only extremely speculative, but, if we may be pardoned for saying so, quite preposterous. We know, for example, that the emperor Humāyūn, and certainly Akbar, the prince royal, took painting lessons from ʿAbd al-Ṣamad[15] much before he became "the head of the atelier." Surely, Khandalavala does not expect us to believe that a teacher good enough for Akbar and his imperial father was not good enough for his talented discovery.

After this extraordinarily feeble argument on the date of Dasavanta's career, Khandalavala concludes therefrom that the *Ṭūṭī-nāma* was painted in ca. 1576–80, but then immediately proceeds to contradict the premise on which he bases his conclusions by stating that the ascriptions in the manuscript "seem to be later additions," and that the manuscript itself was prepared for a patron other than the emperor. Now the painting is either by Dasavanta or it is not. If it is, then by Khandalavala's own arguments, it must be a royal copy, it being inconceivable for a painter who could be trained by none other than the head of the royal atelier to work for a lesser patron. And, if it is not, then it should cease to be used as an argument.

The most lengthy and ingenious criticisms of our views are contained in a recent article devoted entirely to the subject by Ananda Krishna.[16] The author bases his argument on opinions previously expressed by Khandalavala, Chaghatai, and Gray, adding some further notions of his own. Unfortunately, his views suffer from the same drawbacks as those he follows, and the overall impression one receives is of an attempt to bolster a preconceived opinion by marshalling every conceivable argument in its support, whether serious or trivial, consistent or contradictory. A prominent weakness is the selective bias in the choice of evidence, with a marked tendency to ignore or misinterpret what is inconvenient to the author's theories. Thus, the strong resemblances between the *Ṭūṭī-nāma* and the *Ḥamza-nāma* are largely ignored, a peculiarity of the *Ṭūṭī-nāma* that should be obvious to any unprejudiced student and which has been demonstrated at some length in the critical

13. For a criticism see *supra*, pp. 66–67, fn. 49.
14. See Appendix B, p. 175.
15. See *supra*, p. 22.
16. Ananda Krishna, "A Re-assessment," pp. 241–68.

descriptions given here. The argument is also generally vitiated by a tendency to draw from the evidence only those conclusions which support his ideas, blithely ignoring those that do not. To give an example, when trees similar to the Chawand *Rāgamālā* series dated 1605 are found in the *Ṭūṭī-nāma,* it is an indication of its late date; when pavilion and sky similar to those in earlier paintings like the Mandu *Kalpasūtra* of 1439 and the 1540 *Mahāpurāṇa* are observed, the possibility of an earlier date is ignored.[17]

M. A. Chaghatai was the first to attempt dating the *Ṭūṭī-nāma* on the basis of the *naskh* script in which it was written. According to him, all "official" manuscripts of the Akbar period were generally written in *nastaʿlīq,* with one exception, namely the British Museum *Anwār-i Suhailī* which was in *naskh,* and according to him, painted between 1592 (sic) and 1610.[18] The *Ṭūṭī-nāma* being written in *naskh,* and like the *Anwār-i Suhailī* also possessing "signatures" of artists had therefore to be dated in the period just before that manuscript, namely 1580–90.[19] The shallowness of this argument and its basic irrelevancy, when it is the style of painting that is under discussion, was not realised by Chaghatai, nor was it by Ananda Krishna, who embroiders the argument, and pursues it with even greater vigour. According to him, essentially, as no imperial illustrated manuscripts of the Akbar period in the *naskh* script are known so far, and as the *Ṭūṭī-nāma* is written in *naskh,* it could not possibly belong to the atelier of that emperor, this notwithstanding the use of *naskh* on coins "as they follow earlier traditions."[20] Why manuscripts should not follow earlier traditions is not stated. We are told, nevertheless, that as the *Niʿmat-nāma* painted in Malwa was written in *naskh,* this would indicate that the *Ṭūṭī-nāma* also belonged to the same provincial tradition.[21] What is ignored, however, is that the *Niʿmat-nāma* is of the first decade of the sixteenth century, which would indicate on the basis of his own comparison an earlier date for the *Ṭūṭī-nāma.* Now, actually, there is no evidence that *naskh* was discarded by Akbar; true, he seems to have preferred *nastaʿlīq,* but *naskh* continued to be used on coins; on the architectural inscriptions of the period, notably one in the tomb of Shaikh Salīm Chishtī at Fatehpur-Sikri dated A. H. 979/1571–2, recording praises of the Shaikh and his death;[22] in numerous unillustrated manuscripts; and even in the illustrated imperial manuscript of the British Museum *Anwār-i Suhailī* referred to by Chaghatai, begun towards the end of the reign and completed in the early years of Jahāngīr. One

17. *Ibid.,* pp. 248, 257.
18. Internal evidence indicates that a more likely date is between A. H. 1013/ 1604–05 (dates on two miniatures) and A. H. 1019/ 1610–11 (the date of the colophon).
19. *Pakistan Times,* 22 November 1964, p. 2, column 6.
20. Ananda Krishna, "A Reassessment," p. 264 and fn. 77.
21. *Ibid.*
22. Husain, p. 66.

might as well argue that this manuscript was a provincial work because of the *naskh*, or that the Cleveland *Ṭūṭī-nāma*, being written in *naskh* similar to that of the *Niʿmat-nāma*, was of the same date; or that it was at least of the pre-Akbar period because Akbar was particularly fond of *nastaʿlīq;* but all these arguments would be as patently without foundation as are his. There is simply no reason as to why there should not have been illustrated imperial manuscripts written in *naskh* in the reign of Akbar, and the possiblility that a manuscript of this type would be early in the reign, when new preferences had not gained currency, would be perfectly feasible.

Ananda Krishna tackles the problem of the ascriptions to painters in the *Ṭūṭī-nāma* in an equally loose manner. According to him, the early imperial manuscripts produced before 1575, namely the *Ḥamza-nāma*, the 1568 *ʿĀshiqa*, the Rampur *Astrology* manuscript, and the 1570 *Anwār-i Suhailī* "fail to mention the names of illustrators;"[23] but as the *Ṭūṭī-nāma* does carry ascriptions, it has to be excluded from this class of early manuscripts characterised by an absence of inscriptions. Moreover, he continues, when inscriptions first begin to be recorded in the "second phase" of Akbar painting, they refer not to single but to two or more painters working on a miniature. As the *Ṭūṭī-nāma* has ascriptions only to single painters, it could not, according to him, belong to this second phase either. He notices that the *Dārāb-nāma,* dated by him to ca. 1580, does have ascriptions to single painters,[24] but instead of recognising the serious difficulties which this presents for his line of reasoning, he proceeds as though the evidence were of little consequence, when in fact it demolishes his argument altogether.

As a matter of fact, there is considerable evidence to indicate that it was the general practice for Akbar period manuscripts and paintings to carry ascriptions to painters, surviving examples running into the hundreds, though of course there may have been exceptions.[25] It is therefore unreasonable to assume, as Ananda Krishna seems to do, that if no ascription is to be presently found on an illustrated folio or a separate miniature, it never was there. Hardly any Akbar period manuscripts have survived in the same state as when they were first made, untouched by the hands of later binders, repairers, and others; and when they have, the ascriptions are invariably present.[26] Only if the manuscripts on which Ananda Krishna bases his arguments had remained untampered in the course of their history, would he have a case; but this is just not so. The area outside the framing lines of the *Ḥamza-nāma*, for example, has been clearly repaired rf. *Ḥamza,* Pls. 1, 14 et al., and the two illustra-

23. Ananda Krishna, "A Reassessment," p. 267.
24. *Ibid.*
25. See *supra,* pp. 56 f.
26. For example the Baltimore *Khamsa* of Niẓāmī. It is a very well-preserved manuscript, even possessing the original book-covers, and has all ascriptions intact.

tions of the *ᶜĀshiqa* are very closely cropped, there being very little margin left, particularly at the bottom. Even a cursory glance at the Rampur *Astrology* reveals that the present format is of a late date, the thin strips of poorly coloured paper around the small paintings and the frames around the larger ones being provided at the time of much later remounting. The 1570 *Anwār-i Suhailī* has been also clearly rebound. There is thus very little basis for postulating a whole class of early Akbar period illustrated manuscripts characterised by the absence of ascriptions, and then denying a manuscript a place in that class because it has some ascriptions. In any case, one is tempted to ask where is the relevance of all this argument when Ananda Krishna has already rejected the ascriptions in the Cleveland *Ṭūṭī-nāma* as spurious, in which case the manuscript would belong to the class of early Mughal ascriptionless manuscripts which he posits![27]

Ananda Krishna's arguments on Dasavanta and the dating of the *Ṭūṭī-nāma* are not different in kind from Khandalavala's, and are perhaps even more rarefied, so much so that we find him a little difficult to follow. For some reason, he feels that the *Āʾīn-i Akbarī* implies that Dasavanta "entered the imperial service probably after the first stages of evolution,"[28] but how these conclusions flow from that work escapes me completely. All that Abū al-Faẓl states, either according to Blochmann's somewhat faulty translation or the new rendering by Naim (Appendix C), is that Dasavanta was discovered by the emperor who had him trained by ᶜAbd al-Ṣamad, and that in a short time he "surpassed all painters," or, more literally, "had no peers." There is absolutely nothing that would interdict a situation in which, for example, the above mentioned events, namely the discovery of Dasavanta, his training by ᶜAbd al-Ṣamad, and his rise to peerless excellence, occurred in the first few years of the reign. To assert otherwise is "reading between the lines" to an excessive and unjustified degree.

In contrast to our views, Ananda Krishna, once again taking his cue from Khandalavala, considers the Cleveland *Ṭūṭī-nāma* to be a provincial product with Mughal influence rather than an early Mughal manuscript in which pre-Akbar elements are being transformed to achieve the Mughal style. Thus, according to him, "it would be unfair to say that the *Ṭūṭī-nāma* artists had even (sic) a direct knowledge of the *Caurapañcāśikā* style," the manuscript being basically one of the *Candāyana* group.[29] As the Mughal *Ḥamza-nāma* has no *Candāyana* features,

27. Ananda Krishna, "A Reassessment," p. 266. The author, on the one hand, writes as though the ascriptions are authentic to allow a late chronological placement of the *Ṭūṭī-nāma* according to his theories; on the other hand the inclusion of Basāvana and Dasavanta among the ascriptions would force him to concede that the manuscript was a royal product, so he decides to disregard them altogether.

28. *Ibid.*, p. 266.

29. *Ibid.*, p. 247. In attempting to summarise Ananda Krishna's views, we have, for the sake of clarity, adapted his designations to those used in this book.

therefore there is no question of the *Ṭūṭī-nāma* having played any part in the development of the *Ḥamza-nāma.* Whatever *Ḥamza-nāma* features are found in the *Ṭūṭī-nāma* have to be explained, according to him, as the result of *Ḥamza-nāma* influence on a *Candāyana* group manuscript of later date. He therefore finds the views of "Chaghatai, Khandalavala, and Gray more acceptable," and prefers to date the *Ṭūṭī-nāma* to ca. 1575–80.[30] As far as the sources of the *Ḥamza-nāma* style are concerned, he sees them not in the *Caurapañcāśikā* and *Candāyana* groups, but in a totally different style, no examples of which are known, but traces of which according to him can be seen in the *Ḥamza-nāma,* and which flourished in the urban environs of Agra.

When each of the above propositions are scrutinised, we find that they are supported by the same kind of weak and defective argument as has been used earlier. It is unnecessary to examine all of them in any detail for their deficiencies and lack of logic are obvious enough, his thesis in any case failing in the light of our own analysis and exposition. All that we propose to do here is to draw attention to the more glaring errors.

Ananda Krishna's assertion that the *Ṭūṭī-nāma* has no *Caurapañcāśikā* features[31] is to be rejected in view of our analysis of any number of its various miniatures.[32] Besides stylistically substantive features, we have pointed out, for example, the common occurrence of a very characteristic mauve and maroon chequer pattern, both in this group and the *Ṭūṭī-nāma.* To this, Ananda Krishna demurs, because, according to him, while the patterns are square in the *Caurapañcāśikā*, they are "diamond" shaped in the *Ṭūṭī-nāma.*[33] Other similarities, including those between the female figures, "the balloon-like *oḍhanī*," and so on are also rejected on similar and extremely trivial grounds.[34] Apparently he has some difficulty in realising that when similar features are found in paintings of two different styles, they are not likely to be exactly the same.

Ananda Krishna also believes that the *Ḥamza-nāma* has no *Caurapañcāśikā* features;[35] what we consider to be *Caurapañcāśikā* features in the manuscript, he does not, again because they are not exactly alike. Rather, he sees similarities to a "local urban style of Agra" which he seems to invent for this particular purpose, and for the existence of which he produces very little evidence. His statement that the

30. *Ibid.*, p. 268.
31. *Ibid.*, p. 247.
32. These have been brought out at some length in our descriptions of the individual miniatures. See particularly *supra*, p. 79, and fols. 43r, 70r, 98v, 99v, 100v.
33. *Ibid.*, p. 248. See however, fol. 93v.
34. *Ibid.*, pp. 247–48.
35. *Ibid.*, p. 244.

Ḥamza-nāma has no *Candāyana* group features is also unacceptable.[36] We have already mentioned some features which he has not noticed, such as the pale ground with sparse tufts of grass (Pl. 2), the deep pink floral pattern, a motif particularly exclusive to the *Candāyana* group (cf. Pl. 29 and Pl. 108), and even striped domes (Pl. 17).[37] No doubt these elements are vestigial, for the simple reason that it is the *Ḥamza-nāma* style that is the dominant one—absorbing and transforming what it takes from pre-Akbar painting, rejecting what is not compatible with its genius.[38] It is a common enough artistic process of which Ananda Krishna seems to be singularly unaware.

His most extraordinary attempts, however, are his efforts to demonstrate the differences between the Cleveland *Ṭūṭī-nāma* and the *Ḥamza-nāma*. He argues, thus, that the work of the *Ṭūṭī-nāma* is more expressive, as is to be seen in the close observation of the donkey's posture as it brays (fol. 207r), "the rolling eyes of the beasts in anger and surprise"[39] (fol. 205v), the irritated monkey (fol. 32v), the vivid grief of the woman lamenting in the graveyard (fol. 10v), and the sensitively depicted agony of the lovers stoned to death (fol. 272v). Astonishingly enough, instead of concluding therefrom a close relationship between the *Ṭūṭī-nāma* and the Mughal style of the *Ḥamza-nāma*, he does quite the opposite, asserting that these features of the *Ṭūṭī-nāma* display "a freedom in treatment which a court art could hardly attain."[40] This, to say the least, is a total misunderstanding of the imperial Mughal style. We have yet to come across examples where the Mughal style is anything if not naturalistically expressive, a quality which Ananda Krishna denies. To assert on the contrary that the presence of these features proves the *Ṭūṭī-nāma* to be a provincial work is really quite incredible, for there is not a single provincial school that possesses this kind of naturalism, which is, as a matter of fact, exclusive to the Mughal school.

Attempts to date the *Ṭūṭī-nāma* to a later period than the *Ḥamza-nāma* are equally fallacious. Attention is drawn, for example, to the red background of some of the *Ṭūṭī-nāma* miniatures (fols. 92r and 110v). As this is a feature which occurs, according to Ananda Krishna, in seventeenth century Rajasthani painting, the *Ṭūṭī-nāma* is also of a late date;[41] to which superficial argument, one can respond by saying that the red background also occurs with monotonous frequency in earlier paintings as well, notably those of the Western Indian style and the *Caurapañcāśikā*

36. *Ibid.*, pp. 257–58.
37. See *supra*, pp. 69, 135.
38. See *supra*, p. 69.
39. *Ibid.*, p. 258.
40. *Ibid.*
41. *Ibid.*, p. 260.

group. Some of the faces of the *Ṭūṭī-nāma,* it is true, are "less primitive or schematic or impersonal" than in the *Ḥamza-nāma,* but this is to be explained by taking into account the varying styles of individual artists, and we find close enough parallels in other works of the *Ḥamza-nāma* period, such as the 1568 *ʿĀshiqa* and the 1570 *Anwār-i Suhailī.* On the other hand, the striking similarities between the faces of the *Ṭūṭī-nāma* and the *Ḥamza-nāma* (see for example, fols. 46v, 49v, 105r, 271r and 291v) are entirely ignored, as are those which are even more primitive.[42] Another argument advanced is the familiar one which relies on the "presence of European features," dealt with earlier;[43] while yet another affirms that the treatment of water being tamed in the *Ṭūṭī-nāma,* it is later than the *Ḥamza-nāma* where the water is much more turbulent. This again is not true (cf. fols. 89r and 89v). Indeed, so desperate is the attempt to latch on to any argument that could conceivably prove his point, that Ananda Krishna, as we have pointed out earlier, sees herons (fol. 8r) where there are only patches caused by peeled colour and proceeds to assign a late date to the *Ṭūṭī-nāma* on the basis of these non-existent birds possessing what he calls "the softness of a more *sophisticated* style" (italics his).[44]

The Ṭūṭī-nāma and the origin of early Mughal painting

Given the Timurid tradition of patronage and connoisseurship of the arts, the growth of a school of painting under the Mughals was inevitable. Humāyūn, even when in unfortunate exile in Persia and preoccupied with thoughts of return, was busy recruiting painters; and before long, in the reign of his son Akbar, a distinctive Mughal style came into being. The character of this style was shaped by the patronage of the emperor and the Indian environment with which he identified himself so whole-heartedly, in contrast to the attitude of his father and, more particularly, his grandfather. This is revealed in his marriages to Hindu princesses, alliances with Hindu powers, and an exceptionally liberal religious policy, and in smaller, and in some senses even more meaningful ways, such as his drinking of Ganges water drawn at Hardwar, in his preference for Hindi names, and his fondness for Indian literature.[45] A new mode of thought, such as this, would, of necessity, have a repercussion in Mughal art so closely associated with royal patronage, and we

42. Particularly deceptive is the attempt by Ananda Krishna to make his point by comparing a female head in the *Ḥamza-nāma, ibid.,* p. 261, Fig. H, with those in the *Ṭūṭī-nāma.* Here he deliberately chooses one of the most atypical examples from the *Ḥamza-nāma,* completely disregarding the more generally prevalent types as they would negate his pre-conceived notions.
43. See *supra,* p. 83; also pp. 153–54, and p. 52, fn. 7.
44. *Ibid.,* p. 261.
45. *Āʾīn* I, pp. 58, 96, 110 ff.

see it reflected immediately and clearly in architecture where full scope is once again given to Indian traditions. To a ruler like Akbar, a rarefied atelier staffed by Persian émigrés, of a kind that may have satisfied Humāyūn, would not do. It had to let its roots grow in the Indian soil, admit Indian artists and Indian traditions into its fold, and allow its forms to be leavened by those of India. Only thus would come into existence a new and appropriate style, a natural and easy synthesis of the Indian and the foreign, and conforming to the emperor's ideals. And the emperor himself took a keen interest in the forging of the new style, influencing the Persian masters away from their traditions, transmuting their forms, in the carefully chosen words of Abū al-Faẕl, to a more sublime level and greater depth of spirit. Akbar discovered artists and had them trained, inspected their output on a regular basis, and encouraged them to paint in a manner that would meet his approbation by the influence of his own knowledge of the art and by distributing rewards and increasing salaries on the basis of performance. The fact that he himself had been trained in painting must have put him in rather effective command, enabling him to appreciate both the limitations and possibilities of change.

The new Mughal style came into being in its very first and also its most ambitious product, the great paintings illustrating the *Ḥamza-nāma*. Unique and distinct among the styles of the period, there is no difficulty in understanding it theoretically as the Indian product of Mughal genius. But the problems posed by the historian of art, of how exactly it came to be so, what were the processes that led to its distinctive formulation, remained unanswered. The Safawi style under S͟hāh Ṭahmāsp, though it explained some features, was unable to account for the individuality of the *Ḥamza-nāma* style, which could also not be explained by the Indian styles of the pre-Akbar period, little known at first, but knowledge of which has been rapidly increasing in recent years. These styles were indeed beginning to provide some flashes of insight, some inkling of the alchemical processes under way in the royal atelier, but a vital link was somewhere missing. The question, how did the Safawi style and the pre-Akbar Indian styles precipitate the great Mughal style, in spite of brilliant surmises, was yet to receive a satisfactory answer.

The Cleveland *Ṭūṭī-nāma* provides that answer, allowing us a glimpse of the Mughal style in the making, demonstrating effectively the actual manner in which what was taken from the various pre-Akbar Indian traditions was transformed in a rapid and, what can only be called, a revolutionary manner.

In this manuscript, we see a fairly large number of artists, identified by name and by hand, largely belonging to the several pre-Akbar traditions, at work in the imperial atelier, and attempting with varying degrees of success to attain the new norms set up by the emperor and the principal masters. For some, the adjustment was a slow process (Painter A), while others were well on the way to success

(Painter U). Not only are the painters of the *Ṭūṭī-nāma* submitting to the discipline and demands of the new style, but their own several traditions have been thrown into a state of flux by this new proximity to fellow painters belonging to traditions other than their own. We thus begin to see a process of amalgamation in which artists of the *Candāyana* group, the *Caurapañcāśikā* group and the Indo-Persian group are unable to retain the distinctive integrity of their own modes of expression as they had been able to do prior to coming to the court and when they worked within the confines of their own traditions. Instead, they commence to give and take from each other, natural enough for painters working side by side. Thus, in a miniature, the general composition and draughtsmanship may indicate one particular group, but we may get, in a facial type or a touch of colour, features that indicate another. In fol. 112r, for example, the lady and the red ground are derived from *Caurapañcāśikā* group traditions, the Indo-Persian group being the source of the circular floral arabesque drawn on this red ground, while the rest of the colouring, as well as several stylistic devices have affinities to the *Candāyana* group. Fol. 122v again shows features equally derived from the *Candāyana* as well as the *Caurapañcāśikā* groups. In instances like these, it can become quite difficult to determine to which particular style the painter of an individual miniature originally belonged. Nor is this all, for on occasion, we find the same artist predominantly using elements from one group in one painting, but relying for elements from another group in another (cf. fols. 43r and 44v by Painter D). What all this demonstrates fairly clearly, as we have shown earlier, is that the *Ṭūṭī-nāma* represents a kind of melting pot in which were present constituents from the various types of paintings then known in India. The artists who painted the manuscript were experimenting with a great variety of styles, forms, colours, patterns, indeed, the cumulative artistic experience of their time, as they worked their way to a new idiom.

A study of the Cleveland *Ṭūṭī-nāma* reveals clearly the existence in it of numerous pre-Akbar characteristics all in a state of change and ferment, but there is also imposed on them a unifying artistic vision resulting in the first emergence of stylistic concepts that we associate with the Mughal style. If pre-Akbar vestiges are stronger here than in any other work of the Mughal school, it is also true that there is not a single painting in which the movement towards the Mughal style of the early Akbar period as represented by the *Ḥamza-nāma,* the 1568 *ʿĀshiqa,* the 1570 *Anwār-i Suhailī,* and the Rampur *Astrology* manuscript is not present. Sometimes the Mughal style is almost fully achieved as we see in the paintings by Dasavanta (fols. 32r and 32v) and Basāvana (fols. 33v, 35r, and 36v), or the work of Painter U (fol. 225r et al.); and sometimes the artist, like Painter J, still has a long way to go (fols. 98v, 99v, and 100v); and in some instances the same painter achieves the Mughal style to a greater extent in one painting than another (cf. fols. 62r and 62v by Painter E; 282v and 287v by Painter BB); but even in the work of painters who are slowest to learn, some clear Mughal features are always present, particularly in the treatment of the human figures, attempts to give depth to the composition, and the modelling of forms. Taking the *Ṭūṭī-nāma* as a whole, we can almost chart out, step by step, this urge toward Mughal expression, from a phase close to pre-Akbar origins to one which is indistinguishable from the fully evolved Mughal style proper. We see this demonstrated, for example, in fols. 43r, 282v, 124r, 102v, 223v, and 316r, taken in that order, and particularly clearly because they all have the same theme, *Khujasta addressing the parrot* as she prepares to set out on her secret assignation. Fols. 43r, 282v, and 124r are still very close to the *Caurapancāśikā* and the *Candāyana* groups, though some Mughal features are present in each one of them, particularly in attempts to model the figures of the woman and the parrot. In paintings like fol. 102v and related examples, the Mughal features become progressively stronger. Depth is given to the architecture, the line is more relaxed, the clothing is endowed with texture, and ornamental patterns are of the type seen in the *Ḥamza-nāma,* as in the inlaid door and the mauve tiles of the floor. Memories of earlier styles, however, are not yet forgotten and continue to be seen in the bed with flowers arranged at the bolster, though here also the bed is given depth and the flowers are placed more naturalistically. In fol. 223v, the transition has been made to the Mughal style proper, the work anticipating the delicacy of de luxe manuscripts like the 1568 *ʿĀshiqa* or the 1570 *Anwār-i Suhailī* (cf. Pls. 35 and 39), while the bold rendering of fol. 316r gives it the appearance of a detail lifted from the *Ḥamza-nāma* (Pls. 4 and 24). Fols. 150r, 62v, 146r, and 152r all possess a composition which is derived from the *Candāyana* group (Pl. 111), the scene being laid in an open landscape with a high curving horizon and a large leaning tree to the side. Of these, fol. 150r has the least Mughal features, barely observable except in the rocks at the corner and those on the horizon's edge, in a slight attempt to model the lower garment of the Brahman, and in some delicate brushwork. In fol. 62v, the colours take on a richer, more Mughal tone, the ground being painted a rich green with fairly luxuriant grass. The trunk of the tree is modelled, and its leaves freely daubed, a feature not to be seen in fol. 150v, or in any of the paintings of the *Candāyana* group. In fol. 146r, the tree, the rocks with a gushing stream of water, the sky with streaked clouds, and the human figures carrying the shrouded lady are more emphatically Mughal in aspect, while in fol. 152r, the rendering of the tree with its lively cluster of yellow-edged leaves and the expressive figures of the animals are indistinguishable from what is seen in the *Ḥamza-nāma.* In none of these pictures, however, has the transition to the full Mughal style been achieved, as we have just seen in fols. 223v and 316r.

This evolution from the pre-Akbar towards the Mughal idiom in the *Ṭūṭī-nāma* is also to be observed, and perhaps more strikingly, in the rendering of specific motifs.

Take for example the popular blossoming mango tree with its drooping leaves and freshly sprouting pink leaves, a motif ever popular in Indian painting, occurring as early as the Mandu *Kālakācārya-kathā* of ca. 1440 (Pl. 67) and found frequently in the *Caurapañcāśikā* group, where its earliest occurrence is in the 1516 *Mahābhārata* (Pl. 73). It is found in almost the same form in fol. 63v, with slender trunk and compact ovaloid top. Each cluster of leaves is ornamentally conceived, with drooping green leaves at the base, and pink shoots and flowers on top. In fol. 42r, this tree is still essentially the same, though presenting a more naturalistic aspect by virtue of the heavy trunk and larger size. In fol. 110v, however, the trunk is modelled and an attempt made to capture the texture. The foliage, also, is more freely disposed, and instead of the strict arrangement in ornamental clumps, we get leaves issuing from separate twigs and diminishing in size towards the extremities. Nor are the fresh shoots uniformly pink, but painted naturalistically in subtle shades of the same colour and often flecked with green. The compact ovaloid shape of the tree-top is broken up by leaves that stray out of its boundaries, and the naturalism is further enhanced by a keen consciousness of the weight of leaves as they bend each twig. The transition to a Mughal idiom, already achieved here, is carried to perfection in the splendid renderings of the mango tree by Dasavanta (fol. 32r) and Basāvana (fol. 35r).

The same progression is also to be seen in the depiction of water. It is sometimes represented in the archaistic basket-pattern (fols. 57v and 120r), but in examples like fol. 135v, this pattern begins to break up, the individual units being interconnected by ripples of water, a stage of development also seen in the *Ḥamza-nāma* (Pl. 10), to finally achieve the full swirling treatment as seen in fols. 89r and 212v. The representation of water in fols. 306r, 307r, and 308v has not fully attained this freedom, but clearly depicts the artist's struggles and attempts in this direction. The pale ground interspersed with dots, high curving horizon edged with a motif consisting of dots and dashes and strips of sky in blue and gold as seen in fol. 20r are all immediately derived from the *Candāyana* and Indo-Persian groups. In fol. 16r, the ground is shaded with clumps of soft green and yellow grass, but the strips of sky are retained; while in fol. 26v, an attempt is made to soften their hard geometrical appearance by obscuring the horizon with the foliage of trees and the strips of gold and blue sky with wisps of cloud. In miniatures like fol. 42r, the curving horizon is decorated with low shaded mounds, while in paintings like fols. 212v and 226r, the rocks are done in a fully Mughal manner, indistinguishable from similar renderings in the *Ḥamza-nāma* (Pls. 18 and 20). These constant struggles and attempts, of which many more instances could be given, make the study of the *Ṭūṭī-nāma* an extremely fascinating and human one, placing us in a position from where we can, as it were, watch the creation of the Mughal style over the shoulders of the artists themselves. It is these struggles and efforts which account for the corrections, the reworking of what must have been unsatisfactory miniatures, that make the *Ṭūṭī-nāma,* and to a much greater extent than the *Ḥamza-nāma,* the training ground and the birthplace of the Mughal style.

The conclusion is thus irresistible that the Cleveland *Ṭūṭī-nāma* belongs to the formative phases of Mughal painting. Apart from the question of its precise date, it is stylistically symptomatic of the kind of work done at the Mughal atelier of Akbar at its very inception, and as such it is perfectly proper to raise the possibility of its being earlier than the *Ḥamza-nāma.* Further consideration, however, entails difficulties, for the simple reason that we do not know enough about the *Ḥamza-nāma,* the various phases of its style, the parts that belong to these phases, and particularly what its style was during the early part of the fifteen years which it took to complete. In view of this, the possibility of the *Ṭūṭī-nāma* being painted at a point in time when work on the *Ḥamza-nāma* had not progressed beyond the early stages would have to be considered, and seems to us to be also plausible. The *Ṭūṭī-nāma* could then be understood as representing a phase in the development of the Mughal atelier when it was expanding rapidly, and new Indian recruits were being led by their fellow artists, as well as the early masters, to achieve the norms of the Mughal style as it was then being expressed in the *Ḥamza-nāma,* a work executed at a generally higher and more ambitious level of performance. From this point of view, the *Ṭūṭī-nama* could be thought of as a rehearsal for the *Ḥamza-nāma,* even if it is stylistically representative of an earlier phase of work.

The date of the Ṭūṭī-nāma

These general considerations apart, let us try and see if we can arrive at a closer date for the *Ṭūṭī-nāma,* and thus a clearer understanding of it on the basis of a comparative study with early Mughal painting. The *Ṭūṭī-nāma* has any number of close similarities to the *Ḥamza-nāma,* and also retains pre-Akbar Indian features to a considerably larger degree, so at the very least, it is contemporary to the *Ḥamza-nāma.* The date of the *Ḥamza-nāma,* according to arguments advanced earlier, is probably ca. 1562–77, so that it is perfectly reasonable to assume that the *Ṭūṭī-nama* was painted sometime during these years as well.[46]

Before proceeding further, it is important to recall the three types of work as exemplified in the three types of manuscripts done in the Akbar period and their characteristics.[47] The first type were large works with elaborate illustrations done

46. See *supra,* pp. 65 ff.
47. See *supra,* pp. 58 ff.

jointly by two or more artists in a comparatively bold style with a generally high standard of achievement, of which the *Ḥamza-nāma* with its 1400 paintings is the best, if somewhat extreme, example. The second type of manuscript, almost at the opposite end of the scale, is the de luxe product with few miniatures, each done by an individual master working in an extremely refined manner, such as the 1568 *ʿĀshiqa* (Pls. 35 and 36) and the 1570 *Anwār-ī Suhailī* (Pls. 39–41). The third type, in its generally rapid and bold rendering and relatively large number of miniatures, reminds us of the first, except that the miniatures are less elaborate and the work is not done jointly, but by single painters. Some of these, of course, could be the work of masters, and on occasion, a few paintings may resemble those in the de luxe manuscripts, though generally this is not the case. The Rampur *Astrology* manuscript (Pls. 43 and 44), the Beatty *Ṭūṭī-nama* (Pls. 47–61), and the British Museum *Dārāb-nāma* (Pls. 45 and 46) are representative examples of this type. The Cleveland *Ṭūṭī-nāma,* with its numerous but relatively simple miniatures, executed boldly and rapidly by painters working individually, is clearly a work of this class, though it too has some fairly refined work which has affinities with work in de luxe manuscripts and was done by masters.

Now, if we bear in mind the class of manuscript to which the Cleveland *Ṭūṭī-nāma* belongs, the simple nature of the work, and the numerous painters employed,[48] it would appear to us that a period of approximately a year or so would be sufficient for completing the manuscript, in spite of the large number of paintings. It therefore becomes necessary to narrow down the broad date of ca. 1562–77 suggested on the basis of the *Ḥamza-nāma* in order to make it more meaningful. To do this, it is appropriate to turn first to other manuscripts in the same class as the *Ṭūṭī-nāma,* namely the *Astrology* in the Rampur Library, the Beatty *Ṭūṭī-nāma,* and the British Museum *Dārāb-nāma.* Of these, the *Ṭūṭī-nāma* is clearly earlier than the last two for the simple reason that they obviously do not possess the strong *Ḥamza-nāma* features which the *Ṭūṭī-nāma* does, and can be therefore kept out of consideration.[49] A comparison with the *Astrology* manuscript also reveals the priority of the *Ṭūṭī-nāma,*[50] and if our date of ca. 1570 for the manuscript is correct, then the *Ṭūṭī-nāma* would have to be painted before that time. This is confirmed by a comparison of the more refined miniatures of the *Ṭūṭī-nāma,* notably the work of Painter U, with similiar work in the 1568 *ʿĀshiqa.* If we take for example fol. 225r or other miniatures by Painter U (fols. 209v, 223v, 231v, and 234r) and compare them with the miniatures of the *ʿĀshiqa* (Pls. 35 and 36), we find it possible to establish a close relationship

48. We estimate more than forty-five painters. See *supra,* pp. 78–79.
49. See *supra,* p. 74.
50. See *supra,* pp. 73 f.

between the two. The Cleveland *Ṭūṭī-nāma* miniatures, however, are at an earlier stage of development, retaining archaisms, however faint, which are absent in the *ʿĀshiqa* miniatures, nor do they possess quite the marked degree of virtuosity and refinement that begin to be observed in the 1568 *ʿĀshiqa* for the first time. I would, therefore, tend to date the work of Painter U a few years earlier, around 1565, which would thus give us an approximate date for the entire Cleveland *Ṭūṭī-nāma.* Making allowances for the pre-Mughal features in the Cleveland *Ṭūṭī-nāma* which are entirely absent from the 1568 *ʿĀshiqa* and only vestigial in the *Ḥamza-nama,* it would appear to us that a date ca. 1560–65 would be the most appropriate. This also allows for a date in the period immediately preceding the *Ḥamza-nāma,* which, as we have pointed out earlier, is not entirely implausible.

The Ṭūṭī-nāma a work of the royal atelier

The date of ca. 1560–65 suggested for the *Ṭūṭī-nāma* can only be maintained if our perception of the manuscript as a work of the Mughal atelier is correct. It would not be the case if the manuscript was a provincial product belonging to a pre-Mughal school but beginning to be influenced by the Mughal style of the *Ḥamza-nāma* period, in which case, it would have to be dated later. As this position has been taken by some, it is important to consider the matter before we close our account of this important manuscript.

The grounds for considering the *Ṭūṭī-nāma* to be a Mughal work produced at the royal atelier are simple and straightforward. There is the style itself, basically naturalistic, emotionally expressive, and concerned with the plasticity of forms, all distinctive characteristics of the Mughal school. These qualities are not present in equal degree in all the paintings of the manuscript, but there is none in which the attempt to achieve them is not made. The similarities to the work of the *Ḥamza-nāma* period are countless, and we also have here not only the works of the famous masters Basāvana, Dasavanta, and Tārā, but also of lesser known painters such as Banavārī, Ghulām ʿAlī, Lālū, Śravaṇa, and Sūrajū, all of them being artists active in the early and middle period manuscripts like the British Museum *Dārāb-nāma,* the Jaipur *Razm-nāma,* the Patna *Tārīkh,* and the Victoria and Albert *Akbar-nāma.* There is also present in the *Ṭūṭī-nāma* a wide variety of work produced by a large number of painters, ranging from the most conservative to the most progressive, from average to superb. This range of work is a feature characteristic of Mughal manuscripts of the royal atelier, particularly manuscripts of the non-de luxe type, and is evident, not only in the *Ḥamza-nāma,* but to as late a period as the Victoria and Albert *Akbar-nāma* (cf. Pls. 62 and 63). On the other hand, we do not know of a single work in the provincial pre-Akbar styles which has as many painters employed on it. At the most, we may get two or three, or a few more. Also the presence in the *Ṭūṭī-nāma* of not just one pre-Mughal style but several of them, each beginning to influence the other, cannot be satisfactorily explained except in the context of the Mughal atelier, for it was only there that artists from various parts of the empire, such as Kashmir, Lahore, Gwalior, and Gujarat, were brought together enabling the kind of amalgam of styles seen in the *Ṭūṭī-nāma* to take place.

On the other hand, to consider the *Ṭūṭī-nāma* to be a provincial work with Mughal influences presents us with many difficulties that cannot be reasonably explained. For this view to be correct, we would have to start off by postulating a local school quite unlike any of the known pre-Akbar schools, but one which was a mixture not only of features drawn from the *Candāyana* group, but the *Caurapañcāśikā* and the Indo-Persian groups also; and the truth of the matter is that we know of no such school having been in existence. On the contrary, this runs counter to the facts, available evidence indicating that these pre-Akbar schools, apart from a few commonly shared features, zealously maintained their distinct individuality, the colour of the *Caurapañcāśikā* group, for example, being anathema to the *Candāyana* group. Nor is this the only difficulty to be faced. An even more unsurmountable one would be providing an explanation for the startling variety of Mughal influence operating on this imaginary mixed style. It is not the general, undifferentiated aspects of the Mughal style that can be hypothecated to be at work here, the kind of influence to be seen, for example, in the Issarda and Ahmedabad *Bhāgavata-purāṇas* (Pls. 85–90) which belong to a late phase of the *Caurapañcāśikā* group when it came under Mughal influence, or for that matter in paintings like the Chawand *Rāgamālā* (Pl. 94) of the Rajasthani school of Mewar. Rather, the influence itself would have to be of a multifarious type and difficult to account for except by conjecturing not just the influence of the Mughal style as such, but a whole variety of Mughal artists at work in this local center. This is an impossible and topsy-turvy situation, quite contrary to the known manner in which Mughal influence has been exercised on various Indian schools throughout their history, and cannot be accepted.

Even if we were to disregard the hopeless difficulties involved in supposing the *Ṭūṭī-nāma* to be a provincial work with Mughal influence, and grant for the sake of argument that this was indeed so, we should expect analogies with paintings such as the Issarda *Bhāgavata-purāṇa* (Pls. 85 and 86) and the even more instructive Ahmedabad *Bhāgavata-purāṇa* (Pls. 87–90), for these are clearly post-Mughal works of the *Caurapañcāśikā* group, showing Mughal influences derived from the style of the *Hamza-nāma* period in the increased feeling for depth and movement, a less rigid line, more plastically conceived forms, and in details such as the rendering of rocks, water, trees, architecture, and even monsters, all exercised in a clear, orderly and comprehensible manner (cf. Pl. 79 and Pls. 85 and 87).[51] Now the *Ṭūṭī-nāma* provides no analogies whatsoever with works such as these, as it must if it is at all to be considered provincial work with Mughal influence, to whatever pre-Akbar group of paintings it may be hypothecated to belong. Thus the *Ṭūṭī-nāma* miniatures on fols. 89r and 120r are totally different from the Issarda and Ahmedabad *Bhāgavata-purāṇa* paintings with comparable subject matter (Pls. 85 and 87), as are fols. 43r and 209v from other comparable miniatures from the same two series (Pls. 86 and 89). A comparison of *Ṭūṭī-nāma* fol. 110v and Pl. 88 from the Ahmedabad *Bhāgavata-purāṇa* further reinforces the point. The parallelisms that one would expect between the *Ṭūṭī-nāma* and these paintings, if it, the Issarda *Bhāgavata-purāṇa,* and the Ahmedabad *Bhāgavata-purāṇa* were indeed all provincial works under Mughal influence, are just not there and leave us with little choice but to reject that point of view which maintains the *Ṭūṭī-nāma* to be a provincial work under Mughal influence. Rather, the process seen at work in the *Ṭūṭī-nāma* is quite the opposite. It is not a case where a group of paintings belonging to a pre-Akbar style, whether of the *Caurapañcāśikā, Candāyana,* or Indo-Persian groups, is seen coming under the influence of the Mughal style, but one involving a movement in the opposite direction, these styles undergoing a process of transmutation and evolution towards the full-fledged Mughal idiom. The style of the Issarda and the Ahmedabad *Bhāgavata-purāṇas* is not moving in the direction of the Mughal style. Instead it is proceeding smoothly in the direction of succeeding provincial styles best represented by the great schools of Rajasthan. The *Ṭūṭī-nāma,* on the other hand, has no natural successors on that side, its promise finding fulfillment in the virtuosity and splendour of the Mughal school of Akbar which is the true inheritor of its achievements.

51. See *supra,* pp. 39 ff.

APPENDIX A
Bāyazīd's Account of the Painters of Humāyūn

Bāyazīd Beg Turkman, a commander of three hundred[1] was a Turk of the Bīyāt tribe, but a native of Tabriz where he was the neighbour of the famous ʿAlī-Qulī Shaibānī who also joined Humāyūn's service, rising to the high rank of Khān Zamān and a commander of five thousand before his downfall and death in the reign of Akbar. Bāyazīd's career was hardly as meteoric. He met Humāyūn at Shāh Ṭahmāsp's court where he was a servant of the royal *īmām,* and entered service at Mashhad in December 1544. He was present at Kabul till at least 1553, after which we notice him working under Munʿim Khān early in the reign of Akbar.[2] He departed for Mecca in 1578, and in 1584 presented himself before Akbar at Fatehpur-Sikri to be appointed *bakāwal* (steward) in 1586–7. Shortly thereafter, he seems to have suffered a paralytic stroke, but was back at work in 1589–90. The next year he dictated the *Mukhtaṣar,* also known as the *Tazkira-i Humāyūn,* which was completed on 13 June 1591. The account of Humāyūn, being written from memory long after the events, and when Bāyazīd was in feeble health, is somewhat rambling and disconnected, and the chronology is often inaccurate; but he furnishes valuable data from personal experience not given by others.[3]

Of particular interest to us is Bāyazīd's account of the artistic activities at Humāyūn's court in Kabul which he witnessed at first hand. This important testimony has been largely ignored by historians of Mughal painting, though M. A. Chaghatai has made an attempt to use it in his study of Khwāja ʿAbd al-Ṣamad.[4] Professor C. M. Naim's translation of the relevant section based on the text as edited and published by M. Hidayat Hosain[5] is given below. The account begins with the arrival, at Astalaf not far from Kabul, of the embassy of Shāh Ṭahmāsp headed by Walad Beg.[6]

1. *Āʾīn* I, p. 563.
2. *Akbar Nāmā* II, p. 311.
3. The above account is abbreviated from Beveridge, "Memoirs of Bayazid," pp. 296–316.
4. Chaghatai, "Khwajah ʿAbd al-Samad," pp. 157 ff. The translation is inaccurate, the author interpolating explanatory remarks without giving any indication that he is doing so.
5. *Tadhkira,* pp. 65–69.
6. The embassy appears to have arrived shortly after the capture of Kabul, and during the festivities attending the circumcision of Akbar which took place some time in March 1541. See *Akbar Nāmā* II, p. 487.

"When Walad Beg was leaving,[7] His Majesty entrusted to him two *farmāns.* In one he sent for the painters Mīr Sayyid ʿAlī and Mullā ʿAbd al-Ṣamad. The other was directed to Mulla Qutb al-Dīn, father of Qāzī ʿAlī Ba<u>kh</u>s<u>h</u>ī, but he could not come to serve as he was pressed by many other matters. Mīr Sayyid ʿAlī and Mullā ʿAbd al-Ṣamad, however, set out the moment they got the *farmān,* and Mullā Muḥammad <u>Sh</u>irwānī and Mullā Fa<u>kh</u>r, the book-binder, also joined their company. When they reached Kandahar, they sent a petition to His Majesty at Kabul. Since the road from Kandahar to Kabul was not safe, His Majesty ordered <u>Kh</u>wāja Jalāl al-Dīn Maḥmūd of Obih, who had at that time been elevated from the post of *mīr sāmān* to that of *ba<u>kh</u>s<u>h</u>ī beg* and who had a large band of retainers, to escort that party of travelers to Kabul. Further, His Majesty did not wish Mullā Maḥmūd <u>Sh</u>irwānī to come to Kabul since the latter did not know any mathematics – that being what His Majesty desired. Consequently the Mullā stayed behind at Kandahar and joined the service of Bairām <u>Kh</u>ān. Forty days had passed since His Majesty's return from the Balkh campaign[8] when Mīr Sayyid ʿAlī, Mullā ʿAbd al-Ṣamad, and Mullā Fa<u>kh</u>r, in the company of <u>Kh</u>wāja Jalāl al-Dīn Maḥmūd, arrived in Kabul and humbly presented themselves before His Majesty, who honoured them with many favours. At that time among the most notable painters there was one Mullā Dost, who could not give up wine at the time His Majesty did, and who had, without His Majesty's permission, joined the service of Mirzā Kāmrān. He was declared by the connoisseurs to have drawn trees and mountains better than Mānī. And God knows best. . .[9]

"Mīr Sayyid ʿAlī, Mulla ʿAbd al-Ṣamad, the *<u>sh</u>īrīn-qalam,* and Fa<u>kh</u>r, the book-binder, presented for His Majesty's sublime viewing the things that they had in those days made or crafted. His Majesty found it desirable to send whatever these people had submitted before him to Nawāb Ra<u>sh</u>īd <u>Kh</u>ān, the ruler of Kashghar, whom he highly regarded but had not met. As for the letter that was sent with the things from Kabul to the <u>Kh</u>ān, it was given by Mullā ʿAbd al-Ṣamad in A. H. 999/ October 1590–91 in Lahore to Bāyazīd, the author of this *Mu<u>kh</u>taṣar.* And the scribe has copied down here that document:[10]

> 'From among those matchless artists who had presented themselves before me in Iraq and Khurasan and were generously rewarded, a group came and joined my

7. The embassy of Walad Beg seems to have taken leave of Humāyūn some time between late 1546 and April 1547 for it passed through Kabul when the city was temporarily held by Kāmrān during this time. See *Akbar Nāmā* II, p. 503.
8. Humāyūn returned to Kabul from the Balkh campaign on 1 Ramāẓan 956/ 23 September 1549, Erskine, p. 375.
9. Here follows an account of perforated poppy seeds, painted rice grains, and a mention of masters in astrology and mathematics, which has not been translated.
10. The letter seems to be a summary and not an exact quotation, lacking the elaborate and elegant forms of address which are unfailingly present.

service in S̲h̲awwāl A. H. 959/September—October 1552.[11] One of them is the painter Mīr Sayyid ʿAlī, the *nādir al-ʿaṣr*,[12] who is matchless in painting *(taṣwīr)*. He has painted on a grain of rice a polo scene — two horsemen stand within the field, a third comes galloping from one corner, while a fourth horseman stands at one end receiving a mallet from a footman; at each end of the field are two goal posts; and at each corner of the rice is written the following couplet:

A whole granary lies within a grain

And an entire world inside a bubble

and at the bottom he has written, 'the humble servant Sayyid ʿAlī, in the month of Rajab A. H. 959/ June-July 1552.

'Another is the painter Maulānā ʿAbd al-Ṣamad, the unique of the time *(farīd al-dahr)*, the *s̲h̲īrīn-qalam*, who has surpassed his contemporaries. He has made on a grain of rice a large field on which a group is playing polo — two posts at one end and two at the other, with seven players on the field and behind them a rank of footmen who hand out mallets, and in the middle of the field a *chob-i qabaq*[13] ... Another of these rare craftsmen is Maulānā Fak̲h̲r the book-binder, who has made twenty-five holes in a poppy seed ... and there is the unique craftsman Ustād Wais, the gold-wire drawer *(zar-kas̲h̲)*, who has made twenty-five gold and silver wires so thin that Mullā Fak̲h̲r could draw them through the holes in the poppy seed. These few things made by these talented people are being sent. Five *(panj?)* pages of the work of the Ethiopians *(kārhā-i zangiyān)*[14] are being sent. A page of ink-drawing *(taṣwīr-i qalam-i siyāh)* by the painter Maulānā Dost, the *nādir al-ʿaṣr*,[15] who has the honour to serve me and who is in painting the Mānī of this time, in gilding unique in the age, and in drafting (?) *(k̲h̲aṭbarī)* and calligraphy quite beyond compare. A page by Maulānā ʿAbd al-Ṣamad, who has painted the day of Nauroz.[16] One page of work each by Maulānā Darwīs̲h̲ Muḥammad and Maulānā Yūsuf. God willing, thousands of works by matchless artists and craftsmen shall be sent later on.' ''

11. As far as Mīr Sayyid ʿAlī and K̲h̲wāja ʿAbd al-Ṣamad are concerned, this date is in conflict with the earlier statement that they arrived forty days after the Balkh campaign. The *Akbar-nāma* also gives 1549 as the date of arrival of the two painters.

12. The title given by Humāyūn to Mīr Sayyid ʿAlī seems to have been *nādir al-mulk* and not *nādir al-ʿaṣr*, if a later interpolation in the *Nafāʾis al-Maʾās̲i̲r* and the inscription on the *Portrait of a young Indian scholar* (Pl. 64) are correct.

13. See Appendix B, fn. 22 for remarks on extraordinary feats of this type.

14. The significance of this statement is unclear. It could perhaps indicate five miniatures of Abyssinian workmanship.

15. The same praise has been conferred on Mīr Sayyid ʿAlī. This would suggest that the words are being used in a general way rather than as a formally conferred title.

16. This could also mean a picture made on Nauroz day. It seems to have been a Mughal custom to make paintings of or on this day. See BWG, p. 148, nos. 232—33 for two paintings by ʿAbd al-Ṣamad made on New Year's day, A. H. 958/ 9 January 1551, and A. H. 965/ 24 October 1557.

APPENDIX B
Mīr ᶜAlā al-Daula and the Date of the Ḥamza-nāma

The *Nafāʾis al-Maʾāsir* ("Riches of Glorious Traditions") by Mir ᶜAlā al-Daula Qazwīnī is an important work dealing with the reign of Akbar. Besides a variety of interesting information, it contains a section giving short accounts of numerous poets, one of whom, writing under the pen-name of Judāʾī, is none other than the celebrated Safawi and Mughal painter Mīr Sayyid ᶜAlī, who was the first to be charged with illustrating the *Qiṣṣa-i Amīr Ḥamza.* A correct interpretation of the information given here would give us a fairly accurate idea of the date of the inception of that work, a vexed but extremely important problem of early Mughal painting.

A critical edition of the text has yet to be published, though work on it is being done by Maulānā Imtiyaz Ali Arshi of the Raza Library, Rampur. Manuscripts are known to be present in a few libraries, but the one chosen here belongs to the Bayerische Staatsbibliothek in Munich.[1] It is certainly the earliest known, and gives some indication of being the autograph copy. More than elsewhere, we find here clear evidence regarding the date of the work, which is of vital importance if its contents are to be of any real use in dating the *Ḥamza-nāma.*

The Munich manuscript (Pers. Codex 3, mts 24 × 14.5 cms) consists of 335 folios, seventeen lines to the folio. It is written in a clear and handsome *nastaᶜlīq* on good rag paper and is almost free of errors. The late binding is of plain black cloth. The author gives his name as ᶜAlā al-Daula b. Yaḥyā al-Saifī al-Ḥasanī, and dedicates his work to Abū al-Muẓaffar Jalāl al-Dīn Akbar Pādis̲h̲āh G̲h̲āzī.

The work itself consists of a *matlaᶜ* containing remarks on the nature of poetry and its various genres; this is followed by twenty-eight *baits*, after which occur the alphabetically arranged biographies of almost 350 poets primarily of the sixteenth century (fols. 1v–271v). There then intervene several folios (272–289) describing events of the years A. H. 980–1 dominated by the Gujarat campaigns. They begin and end abruptly and clearly give the impression of being addenda, the paper, ink and handwriting being very similar to fols. 1v–271v. A *maqtaᶜ* divided into three *maṭlabs* describing the genealogy, the conquests, and the personal attributes and

1. See Storey, vol. I, part II, pp. 800–801; Sprenger, pp. 46–55; Aumer, pp. 2–3; Taraporevala and Marshall, p. 59. In addition to the Munich manuscript we have been able to consult the manuscripts in the Mulla Feroz Library of the Cama Institute, Bombay (no. 147), and the Maulana Azad Library, Aligarh Muslim University. This manuscript is dated 10 Rabīᶜ I 1085/ 14 June 1674.

achievements of Akbar follows on fols. 290r–334v. The conclusion occurs on fol. 335r (Pl. 120); after describing the contents of the book as consisting of marvelous matters, amazing events, attractive verses, surprising anecdotes, information on cities, and descriptions of lives of sultans and divines, the author states that the date of the inception having been given in the title of the work (A. H. 973/ July 1565–6), which was also a chronogram, it was now necessary to give the date of its completion. Then follows the chronogram, *tammat ʿalā yadaihi,* equivalent to A. H. 979/ May 1571–2, immediately, beneath which, the date 979 is also written in Arabic numerals.[2] This is followed by an Arabic inscription in three lines according to which the humble "Niẓām b. Aḥmad Dihlawī wrote some sections of the book on 8 Rabīʿ II A. H. 980/ 18 August 1572 in the city of Ajmer." In the bottom left corner is a mutilated marginal note according to which a new chronogram had been composed as further conquests of the emperor and other events had occurred which had been added to the book since its completion.[3] Unfortunately, the chronogram itself was cut off when the paper was trimmed for rebinding. On fol. 335v is a note stating that the book was presented to the emperor Akbar who approved of it and desired that a genealogical account be also included.

The Munich copy has a large number of marginal notations[4] which seem to have been written by the author, and these amplify observations in the text or add new information. Thus, the marginal note on fol. 55 (Pl. 119), written towards the end of the biography of Judāʾī (Mīr Sayyid ʿAlī) translates: "At present, the Mīr having obtained permission to go on Haj, the task of preparing the aforementioned book has been assigned to the matchless master K͟hwāja ʿAbd al-Ṣamad, the painter from Shiraz; the K͟hwāja has greatly endeavoured to bring the work to completion, and has also notably reduced the expenditure." On fol. 333r the marginal note translates: "At the king's command, from Agra to Ajmer, at every *kuroh* (two miles), a tower has been made with the horns of the deer bagged in the hunt; this work now being near to completion." The task was in fact completed in A. H. 981. No marginal note, as far as we can determine, relates to an event later than A. H. 981. From this it would appear that these could have been written sometime between A. H. 979, the date of the completion of the work, and A. H. 981. As the work of copying the

2. It is not clear why Aumer, p. 2, chose to interpret this date as 989 unless there is a misprint. The chronogram clearly yields 979.
3. This note, which clarifies the chronology of the manuscript was unfortunately overlooked by Aumer, causing him to draw wrong conclusions regarding the veracity of the dates A. H. 979, 980, given by the author and the scribe respectively.
4. Aumer does not appear to have paid any attention to these notes and makes no mention of them in his account.

manuscript was completed in A. H. 980, the most likely dates for the marginal notes could be narrowed down to A. H. 980–81. The later dates which occur in the margins in no way contradict the date of the text proper of the *Nafā᾿is* which is given by the chronogram as A. H. 979. That is to say, no dates later than A. H. 979 occur in the text proper.

In later copies, the marginal information of the Munich manuscript was included in the text. The entry pertaining to Mīr Sayyid ᶜAlī's pilgrimage and the entrusting of the *Ḥamza-nāma* to Khwāja ᶜAbd al-Ṣamad, for example, forms part of the text proper in the Bombay and the Aligarh manuscripts. The Aligarh manuscript also includes the information regarding the horned milestones and adds a chronogram, *mīl-i shākh* = A. H. 981. This process accounts for the apparent contradiction, in manuscripts later than the Munich copy, between the date of the book as given in the colophon and the material which it contains. Some manuscripts like the one described by Sprenger, have even later dates such as A. H. 991 and 996, which refer to the death of Muḥammad Ḥusain Ḥāfiẓ (A. H. 991/ January 1583–4) and Abū al-Hādī (A. H. 996/December 1587–8).[5] These can be easily understood as information interpolated into manuscripts from a desire to bring accounts upto date when new copies were prepared; and they again in no way affect the date of the main text as found in the Munich manuscript which is to be firmly placed between A. H. 973 and 979.

One apparent difficulty still remains, namely the account of the years A. H. 980–81 which occurs on fols. 272–79 of the Munich copy.[6] The events recounted therein begin with Akbar setting out from Fatehpur-Sikri for Ajmer on 20 Ṣafar 980/ 4 July 1572, and include the birth of Dāniyāl, the defeat of Ibrāhīm Ḥusain Mirzā at Sarnal and his flight, the capture of the city of Surat, and the beginning of the second rapid march to Gujarat after its conquest on 24 Rabīᶜ II 981/ 30 August 1573. These, it is true, seem to contradict the date of A. H. 979 on which the author clearly states his work was completed. But the difficulty disappears if we recall the mutilated marginal note on the last page (fol. 335r) which refers to accounts of further conquests having been added to the book after its completion, and the consequent composition of a new chronogram, an obvious reference to these folios and the marginalia.

The description of the Gujarat campaign ends somewhat abruptly, just as the author begins to warm up to the subject of the celebrated nine-day march, as though

5. Sprenger, pp. 46, 48, 54. The Oudh copy was earlier than 1660 to judge from the entry of a previous owner's name on that date.

6. Aumer, p. 2, erroneously states that these events extended over the years A. H. 980–85, for this is clearly not the case. Nor is it a fragment from a history of Gujarat, but rather an account of Akbar's reign in A. H. 980–81 which was dominated by occurrences in that province.

he had intended to continue his narration but was for some reason interrupted, the last two lines on fol. 289r being left blank. This, together with the absence of marginalia in this section, indicates that it was the last thing to be written. The author apparently intended to have a new version of his work prepared, including in it the Gujarat campaigns at the appropriate place in the *maqtaᶜ*, and also the information in the marginalia. It was probably in connection with the planning of this new version that he noticed the conflict developing between the earlier chronogram and the additional information and therefore composed a new chronogram resolving these differences.

According to the inscription written after the chronogram, Niẓām scribed only some sections of the *Nafāʾis,* though he does not specify which particular ones. In view of the nature of the manuscript, particularly the marginalia and the addenda, we strongly suspect that it was none other than the author, Mīr ᶜAlā al-Daula himself, who wrote the rest. According to us, the *matlaᶜ*, the *baits,* and the account of the Gujarat campaign are in his hand,[7] while the *maqtaᶜ* was scribed by Niẓām. This is suggested to us by the character of the writing and the spacing, though it is possible that in this we may be deceived. The date and place of Niẓām's work, which is 8 Rabīᶜ II 980/ 18 August 1572 at Ajmer is also suggestive. We conjecture that the author was resident at Ajmer writing the Nafāʾis in his own hand when he was interrupted by the arrival of the emperor at that city on 15 Rabiᶜ I 980/ 15 July 1572. He thereupon instructed Niẓām to complete the manuscript, which was achieved on 8 Rabīᶜ II 980/ 18 August 1572, in time for it to be presented for the emperor's inspection before he quitted Ajmer for Gujarat on 22 Rabīᶜ II/1 Semptember 1572. The postscript on fol. 335v according to which the manuscript was seen by the emperor confirms this suggestion. Subsequently Mīr ᶜAlā al-Daula seems to have revised the manuscript, making marginal notes. He also decided to add events which had occurred since he had first completed the work in A. H. 979, and wrote out the important occurrences from A. H. 980 to 11 Muḥarram 981/15 May 1573 on which day began the famous nine-day forced march to Gujarat. The account in the Munich manuscript is incomplete, the last two lines of the folio being left blank.

Whether our conjectural reconstruction of the circumstances in which the Munich copy of the *Nafāʾis* was written is correct or not, two points are abundantly clear. Firstly, the portion of the text dealing with Mīr Sayyid ᶜAlī Judāʾī was written, without any doubt, between A. H. 973 and A. H. 979/ 29 July 1565–25 May 1571. Secondly, the marginal note referring to the Mīr having proceeded on pilgrimage

7. Mīr ᶜAlā al-Daula refers to his own accomplishments as a calligrapher in the *Nafāʾis.*

and the *Ḥamza-nāma* project being supervised by Khwāja ʿAbd al-Ṣamad was written after A. H. 979 when the manuscript was completed, and most probably during A. H. 980 and A. H. 981/ 14 May 1572-22 April 1574 when the marginal entries were made.

Of the life of Mīr ʿAlā al-Daula we know little except what we can glean from the *Nafāʾis* and Badāyūnī.[8] He belonged to a distinguished and learned family, particularly known for its great knowledge of history. His father, Mīr Yaḥyā, was a celebrated historian, being the author of the *Lubb al-Tawārīkh,* a general history written for Bahrām Mirzā, brother of Shāh Ṭahmāsp, which covered events from the earliest times to the date of its composition, A. H. 948/ 1541–2. Mīr Yaḥyā was a favourite of Shāh Ṭahmāsp, but fell into disfavour because of his Sunnī faith, and was imprisoned in Isfahan where he died about A. H. 962/1554–5. Mīr ʿAbd al-Laṭīf, the elder brother of Mīr ʿAlā al-Daula, fled to India, where he was received with honour and appointed tutor to the young Akbar in the second year of his reign and soon became the young emperor's confidant. He died at Fatehpur-Sikri on 5 Rajab A. H. 981/31 October 1573 and was buried at Ajmer. His son, Ghiyās al-Dīn ʿAlī, alias Naqīb Khān, Mīr ʿAlā al-Daula's nephew was a boon companion of the emperor Akbar and renowned for his goodness and scholarship, particularly in history. He died in the reign of Jahāngīr in A. H. 1023/ 1614–15.

Mīr ʿAlā al-Daula appears to have followed his distinguished brother to India after a few years, arriving there sometime between A. H. 971/ August 1563–4 when he is known to have been still at Qazwin and A. H. 973/July 1565–6 when we find him at Lahore, the very year in which he began writing the *Nafāʾis.* He was with the emperor at the siege of Chittor in A. H. 975/1567–8. On 8 Rabīʿ II 980/ 18 August 1572, he seems to have been at Ajmer where some parts of the Munich manuscript were copied, a period coinciding with the emperor's halt at the city on his way to Gujarat. He apparently accompanied the emperor on the first Gujarat expedition for he was with the Khān-i Aʿẓam Mirzā ʿAzīz Koka when the latter was appointed governor of Gujarat on 15 Shaʿbān 980/21 December 1572, composing a chronogram for the occasion. We next see him with the emperor at Agra sometime in A. H. 981/ 1574, addressing a letter on the emperor's orders to Munʿim Khān, Khān-i Khānān, who was campaigning in Bihar. The last we hear of Mir ʿAla al-Daula is early in A. H. 983/ 1575 when ʿAbd al-Raḥīm Mirzā Khān was appointed

8. Much of the information given in the *Maāthir-ul-umarā* under Naqīb Khān, vol. II, pp. 381 ff. has been taken from the *Nafāʾis.* Badāyūnī III, pp. 148 ff. gives some additional details in his account of Mīr ʿAbd al-Laṭīf, but has little to say about ʿAlā al-Daula or his poetry, see p. 437. Also see Fakhruzzaman, "Mir ʿAlaud-Daulah," pp. 31–48.

to be the nominal governor of Gujarat, and he was named to accompany him as *amīn.*[9]

As Mīr ʿAlā al-Daula himself states, he was interested in poets and poetry from a very early age, making enquiries about them and their work, committing some facts and verses to memory and keeping records of others. He was himself a poet, writing under the pen-name of Kāmī, but his chief claim to fame is as the author of the Nafāʾis which has been praised for its comprehensiveness, wide range of learning, and critical acumen.[10] It formed the basis of Badāyūnī's account of the poets of Akbar's reign.[11] The Mīr was also much attracted to the art of calligraphy, having learned it from one Mīr Kulangi. His scholarly temperament, and the proximity of both himself and his family to the person of the emperor lend the greatest reliability to his work.

A translation of the account of Mir Sayyid ʿAlī Judāʾī in the manuscript of the Bayerische Staatsbibliothek, Munich, fols. 54r–55r (Pls. 117–119) by Professor C. M. Naim of the University of Chicago is given below. Grateful acknowledgements are expressed to Professor Momin Mohiuddin of the University of Bombay; Dr. Z. A. Desai of the Archaeological Survey of India; Maulana Imtiyaz Ali Arshi and Akbar Ali Khan Arshizadah of the Raza Library, Rampur; Professor Heshmat Moayyad of The University of Chicago; and Mr. Charles W. Ervin who have all helped in one way or another in studying the complex problems involved.

"Judāʾī: His name is Mīr Sayyid ʿAlī and he is a true successor to his father Mīr Muṣawwir. His origin is from Tirmiz,[12] though at various times his ancestors have lived at Badakhshan. The Mīr possesses a great many talents. In the field of painting, as it is his hereditary art, he is a matchless master; in the composition and discernment of poetry, he is well accomplished and well informed[13] . . . As a child and as a youth he was brought up in Iraq.[14] During A. H. 956/1549 he arrived in Kabul and gained the honour of serving *ḥaẕrat-i jannat-ās͟hiyānī,*[15] who by nature inclined to favour all talents, and specially painting, bestowed on him much favour

9. *Akbar Nāmā* III, p. 236. Arshi, *Tārīk͟h-i Akbarī,* fn. on p. 14 gives the date of his death as A. H. 982, but he was alive in the beginning of A. H. 983. Badāyūnī, who completed the *Muntak͟hab al-Tawārīk͟h* in A. H. 1004/1596 does not specifically mention him as being dead.

10. Nizami, "Perisan Literature under Akbar," p. 324.

11. Badāyūnī III, p.239.

12. Both the Bombay and the Aligarh manuscripts also state his place of origin as being Tirmiz, though the *Āʾīn* (see Appendix C) refers to him as Mīr Sayyid ʿAlī of Tabriz. As Tirmiz is near Badakhshan where his ancestors lived, it is possible that the Mīr lived there before he proceeded to Tabriz, which remained the Safawi capital till 1548 in spite of frequent occupation by the Turks.

13. Here occurs one verse which has not been translated.

14. That is ʿIrāq-i ʿAjamī, roughly Persia beyond Khurasan.

15. "Nestling of paradise," the post-mortem title of Humāyūn.

and attention, and greatly praised his paintings. At present he is the cynosure of the transmuting sight *(kīmiyā āsār)* of His Majesty *(ḥaẓrat-i aʿlā)*[16] . . . He has composed many panegyrics, short poems, and quatrains, all well crafted and well received.

"It is now seven years that the Mīr has been busy in the royal bureau of books *(kitāb khāna-i ʿāli)*, as commanded by His Majesty *(ḥaẓrat-i aʿlā)*,[17] in the decoration and painting of the large compositions *(taṣwīr-i majālis)* of the story of Amīr Ḥamza *(qiṣṣa-i amīr ḥamza)*, and strives to finish that wondrous book which is one of the astonishing novelties that His Majesty has conceived of. Verily it is a book the like of which no connoisseur has seen since the azure sheets of the heavens were decorated with brilliant stars, nor has the hand of destiny inscribed such a book on the tablet of the imagination since the discs of the celestial sphere gained beauty and glamour with the appearance of the moon and the sun. His Majesty has conceived of this wondrous book on the following lines. The amazing descriptions and the strange events of that story are being drawn on the sheets for illustrations in minuscule detail and not the subtlest requirement of the art of painting goes unfulfilled. That story will be completed in twelve volumes, each volume consisting of one hundred leaves *(waraq)*; each leaf being one 'yard' *(zarʿ)* by one 'yard,' containing two large compositions *(majlis-i taṣwīr)*.[18] Opposite each illustration, the events and incidents relative to it, put into contemporary language, have been written down in a delightful style. The composition of these tales, which are full of delight and whet your fancy, is being accomplished by Khwāja ʿAṭāullāh, the master prose stylist *(munshī)* from Qazwin.[19] Although, during the aforesaid period, thirty painters,[20] equal to Mānī and Bihzād, have constantly been devoted to the task, no more than four volumes have been completed.[21] One can imagine just from that its grandeur and perfection. May God bring their work to completion under the sublime and majestic shade.

16. The title *ḥaẓrat-i aʿlā* refers to the reigning monarch, in this case Akbar, and not Humāyūn who is referred to but a few lines earlier by his post-mortem title.
    Here follow examples of Mīr Sayyid ʿAli's verses which have not been translated. The Aligarh manuscript interpolates a sentence which translates: "And he is distinguished by the high title of *nādir al-mulk-i humāyūn shāh.*" The sentence is not found in either the Munich or the Bombay manuscripts.
17. The testimony, as indicated by the use of the title *ḥaẓrat-i aʿlā*, is clear and unequivocal. Chaghatai's assertion, "Mir Sayyid Ali," p. 26, also made on the strength of ʿAla al-Daula, that the *Ḥamza-nāma* was begun under Humāyūn is clearly wrong and appears to be based either on a faulty manuscript of the *Nafāʾis* or an erroneous interpretation.
18. Literally "group pictures." The words have been rendered by Arnold as "large compositions." See BWG, p. 190.
19. Khwāja ʿAṭāullāh is the author of the new version, not its scribe.
20. In the Bombay manuscript, the scribe has accidentally omitted the number of painters. The Aligarh manuscript states that these were fifty.
21. The word *chahār* (four) seems to be superimposed over another word which appears to have been *do* (two). Both the Bombay and the Aligarh manuscripts read *chahār*.

"At present, the Mīr having obtained permission to go on Ḥaj, the task of preparing the afore-mentioned book has been assigned to the matchless master Khwāja ʿAbd al-Ṣamad, the painter from Shiraz; the Khwāja has greatly endeavoured to bring the work to completion and has also notably reduced the expenditure."[22]

22. This last paragraph does not occur in the main text but has been added on the margin in a firm and neat hand. The Bombay manuscript and the Aligarh manuscript which is dated 1674 include this sentence in the text proper. The Aligarh manuscript also goes on to state: "And that Khwāja, during the period of Humāyūn, did rare things in the realm of painting such that imagination fails to conceive of them. One was that on the face of a single grain of rice he painted a polo field with riders and men on foot such that the colour of each horse and the face of the person were quite distinct." I do not quite know what to make of references to these extraordinary objects, though here the statement is clearly a later interpolation. We find even more elaborately decorated grains of rice attributed by Bāyazīd to both Mīr Sayyid ʿAlī and Khwāja ʿAbd al-Ṣamad (Appendix A). What I suspect is that these miraculously executed objects became part of the stock description of famous painters and were given free currency in later times, but I am not certain of this. Mirzā Ḥaidar, according to the manuscript used by Arnold, also refers to Darwīsh Muḥammad, his teacher, painting a man on horseback lifting a lion on the point of a javelin, all painted on a grain of rice. See BWG, p. 191.

APPENDIX C
Abū al-Faẓl's Account of Mughal Painting

The *Āʾīn-i Akbarī* of Abū al-Faẓl, which is the third volume of the *Akbar-nāma,* was completed, according to Blochmann, its editor and translator, in the forty-second year of Akbar's reign (1597–8), having taken seven years to complete, a slight addition being made in the forty-third year (1598–9) on account of the conquest of Berar.[1] The section on painting occurs in *āʾīn* 34, entitled *Āʾīn-i taṣwīr-khāna.*[2] The text, though brief, is of the greatest importance for the understanding of Mughal painting. The skillful exposition, employing a precise technical terminology, reveals Abū al-Faẓl to be a connoisseur in the tradition of Mirzā Muḥammad Ḥaidar, Bābar's cousin, who wrote a fine account of the painters of Herat.[3] As doubts have arisen regarding the accuracy of Blochmann's translation, I requested Professor C. M. Naim of the University of Chicago to prepare a fresh translation keeping in mind the problems of a student of Mughal painting. This translation, which is based on the text as edited by Blochmann, is given below.[4]

"Drawing the likeness *(shabīh)* of anything is called *taṣwīr* (painting, pictorializing). Since it is an excellent source, both of study and entertainment, His Majesty, from the time he came to an awareness of things,[5] has taken a deep interest in painting and sought its spread and development. Consequently this magical art has gained in beauty. A very large number of painters has been set to work.[6] Each week the several *dāroghas* and *bitikchīs*[7] submit before the king the work done by each artist, and His Majesty gives a reward and increases the monthly salaries according to the excellence displayed. His Majesty has looked deeply into the matter of raw

1. Biography of Abū al-Faẓl in his translation of the *Āʾīn-i Akbarī* I, p. liii. The forty-third year is given in the Christian era as 1596–7, which is an error for 1598–9; see *Akbar Nāmā* III, p. 1102, where the year is stated as beginning on 15 Shaʿban 1006/ 11 March 1598. Also see *Āʾīn* III, p. 476.
2. This was rather freely translated by Blochmann as "The Arts of Writing and Painting," probably as being more descriptive of the contents. It is interesting to note the specific reference to the *taṣwīr-khāna* ie. the atelier of painters which testifies to its particular importance. That the chapter is not called *āʾīn-i kitāb-khāna* is also significant.
3. *Supra,* pp. 11 f.
4. *Āʾīn* (text) I, pp. 116-18. In making the translation use has been made by Professor Naim of variant readings given by Blochmann when they make better sense. For the English translation by Blochmann see *Āʾīn* I, pp. 113–15.
5. I.e., since his childhood.
6. *Āʾīn* (text), p. 116, variant 13.
7. Names of officials of the workshop, probably the superintendants *(dāroghas)* and clerks *(bitikchīs).* The latter word means a writer; but he also kept accounts, and performed secretarial duties. See *Āʾīn II,* pp. 50–52.

materials and set a high value on the quality of production *(paidā'ī)*. As a result, colouring has gained a new beauty *(rang-āmezī ḥusn-i dīgar pazīraft)*,[8] and finish a new clarity *(ṣafāhā rā ābrū-i tāza padīd shud)*.[9] Such excellent artists have assembled here that a fine match has been created to the world-renowned unique art of Bihzād and the magic making of the Europeans *(ahl-i farang)*. Delicacy of work *(nāzukī-i kār)*,[10] clarity of line *(safā'ī-i nuqūsh)*, and boldness of execution *(ṣabāt-i dast*, lit., stability of the hand), as well as other fine qualities have reached perfection, and inanimate objects appear to come alive. More than one hundred persons have reached the status of a master and gained fame; and they are numerous who are near to reaching that state or are half-way there. What can I say of India! People[11] had not even conceived of such glories; indeed, few nations in the world display them (such glories).[12]

"Among the forerunners on this high road of knowledge *(āgahī)* is Mīr Sayyid ʿAlī of Tabriz. He had learnt a little from his father. When he obtained the honour to serve His Majesty and thus gained in knowledge, he became renowned in his profession and bountiful in good fortune. Next there is Khwāja ʿAbd al-Ṣamad, the *shīrīn qalam* (lit., sweet pen/brush) of Shiraz. Though he knew this art before he joined the royal service, the transmuting glance *(iksīr-i bīnish)* of the king has raised him to a more sublime level and his images have gained a depth of spirit.[13] Under his tutelage many novices have become masters. Then there was Daswanta, the son of a palanquin-bearer *(kahār)*,[14] who was in the service of this workshop and, urged by a natural desire, used to draw images and designs on walls. One day the far-reaching glance of His Majesty fell on those things and, in its penetrating manner, discerned the spirit of a master working in them. Consequently, His Majesty entrusted him to the Khwāja. In just a short time he became matchless in his time and the most excellent *(sar-āmad-i rūzgār)*, but the darkness of insanity enshrouded the brilliance of his mind and he died, a suicide. He has left several masterpieces. In designing

8. Arnold, *Library of A. Chester Beatty*, I, p. xxv translates the word *rang-āmezī* as "blending colours," and considers it to be synonymous with *ʿamal*, which can be also extended to mean the entire process of painting a picture, p. xxvi.
9. Schroeder, p. 10: "True painting should be *saf*, limpidly clear and clean, naked, sheer as a cliff."
10. Schroeder, p. 10, interprets the word to mean "delicacy, slenderness, vulnerable softness."
11. *Ā'īn* (text), p. 117, variant 4.
12. Cf. Blochmann's translation, *Ā'īn*, p. 114: "This is especially true of the Hindus; their pictures surpass our conception of things. Few, indeed, in the whole world are found equal to them." This rendering is incorrect and misleading. Also see the more accurate translation by Arshi, quoted in Khandalavala and Mittal, "An Early Akbari Illustrated MS.," p. 11.
13. The phrase used is *ṣūrat-i-ū rūī ba-maʿnī āword*, lit., his forms turned their face to deeper meanings.
14. The word is spelled with both a *kāf* and *qāf* in manuscripts. Dr. Moti Chandra has also suggested the reading *gohār*, the second stroke on the flag of *gāf* being commonly omitted at the time. It is the name of a caste that earns its living by making paintings on walls upto the present day.

(ṭarrāḥī),[15] painting faces *(chihra kushā'ī)*, colouring *(rang-āmezī)*, portrait painting *(mānind nigārī)*,[16] and other aspects of this art, Basāwan has come to be uniquely excellent. Many perspicacious connoisseurs give him preference over Daswanta. The other famous and excellent painters are Kīsū, Lāl, Mukund, Miskīn, Farrukh the Qalmāq, Mādho, Jagan, Mahes, Khemkaran, Tāra, Sāwlā, Harbans, and Rām. My discourses would get too long if I were to discuss each of them, so I have chosen just one flower from each garden and just one ear from every sheaf.

"It is indeed amazing that from a cultivation of the habit of observing and making of images–which is by itself a source of indolence–came the elixir of wisdom and a cure for the incurable sickness of ignorance, and those many haters of painting who blindly followed their predecessors had their eyes opened to Reality. One day in an intimate assembly, when the fortunate few had gained his proximity, His Majesty remarked: 'I cannot tolerate those who make the slightest criticism of this art. It seems to me that a painter is better than most in gaining a knowledge of God. Each time he draws a living being he must draw each and every limb of it, but seeing that he cannot bring it to life must perforce give thought to the miracle wrought by the Creator and thus obtain a knowledge of Him.'

"As this art gained in status more and more masterpieces were prepared. Persian books of both prose and poetry were decorated and a great many large and beautiful compositions *(majlis)* were painted. The story of Ḥamza *(qiṣṣa-i ḥamza)*, put into twelve volumes, has been illustrated *(rang-āmez kardand)*, and magic making masters have painted fourteen hundred astonishing pictures of as many incidents *(mauziᶜ)*. The *Chingiz-nāma*, the *Zafar-nāma*, this book, the *Razm-nāma*, the *Rāmāyan*, the *Nal-Daman*, the *Kalīla wa Dimna*, the *ᶜIyār-i Dānish*, and other books have been illustrated *(paikar-nigārī)*, His Majesty himself having indicated the scenes to be painted. At His Majesty's command portraits have been painted of all of His Majesty's servants and a huge album *(kitāb)* has been made. Thus the dead have gained a new life, and the living an eternity. Just as the painters have gained in status so have the illuminators *(naqqāshān)*, gilders *(muzahhibān)*, margin-makers *(jidwal-ārāyān)*, and book binders *(ṣaḥḥāfān)* found themselves to be in great demand. By serving in this workshop a great many *manṣabdārs*, *aḥadīs*, and other troopers have gained in distinction. The pay of a foot-soldier is not more than 1200 *dāms* and not less than 600 *dāms*."

15. Cf. Schroeder, p. 10: "*Tarh* (design), which means diagram, intention; and 'program' comes nearer to it than our words 'sketch' or 'study'." The word may perhaps be best understood in the sense of an architect "designing" a building.

16. Arnold, *Library of A. Chester Beatty*, I, p. xxv translates the word as 'taking likenesses.'

APPENDIX D
Ḥājī Muḥammad ᶜĀrif's Remarks on the Ḥamza-nāma

Ḥājī Muḥammad ᶜĀrif Qandahārī, as his name indicates, was born in Kandahar and is known to have been in the service of Bairām K͟hān, K͟hān-i K͟hānān, after whose death he went on a pilgrimage to Mecca. He subsequently joined Muẓaffar K͟hān Turbatī, and is recorded to have been in his service in A. H. 982/ April 1574–5. Muẓaffar K͟hān was a commander of four thousand and one of the great nobles of Akbar's court, even if his career was stormy and erratic. Its high point was reached with his appointment to the office of *wakīl* in 1573, a post which he held for only a short time, having incurred the displeasure of Akbar. We next find him fighting in Bihar where he so distinguished himself as to be appointed governor in 1575. When Muẓaffar returned to the court in 1577, where he stayed for two years, Muḥammad ᶜĀrif accompanied him. In 1579, Muẓaffar K͟hān was appointed governor of Bengal. He lost his life in 1580, being executed by rebels who had succeeded in capturing the city of Tanda which he was defending.

Muḥammad ᶜĀrif was the author of a work the exact title of which is not known, but which is referred to as the *Tārīk͟h-i Akbarī* or the *Tārīk͟h-i Qandahārī*. It was written for Muẓaffar K͟hān and was intended to be called the *Muẓaffar-nāma;* but after the latter's death in A. H. 988/ 1580, the author dedicated it to the emperor, apparently in an attempt to gain a new patron. The work survives in an extremely fragmentary condition in two manuscripts, both of them very corrupt and full of errors, one in the Cambridge University Library, and the other in the Raza Library, Rampur; these have formed the basis of the edited and published text.[1] According to the editors, the work was modelled after the *Nafāʾis al Maʾās̱ir,* and the author relies heavily on that work for his information also, often using identical language and quoting the same verses. When complete, it should have possessed a *matlaᶜ* as well as a *maqtaᶜ*, but all that is preserved is the *muqaddama* portion of the *maqtaᶜ*.

The Cambridge manuscript seems to have been written in or prior to A. H. 986, as the last event reported there is of that date and also because the present tense is used in recounting the events of that year, for example: "At this time when it is A. H. 986, the king has had the tank emptied of water and filled with copper, silver, and gold coins . . ."[2] In any case the manuscript was written when Muẓaffar K͟hān

1. *Tārīk͟h-ī Akbarī,* Rampur 1962.
2. *Ibid.,* p. 152

was alive, and his name as well as that of the emperor are used in the laudatory phrases. This copy, however, appears to have been revised by the author after the death of Muẕaffar K͟hān, some changes being made, most significantly the crossing out of Muẕaffar's name, apparently in preparation for a re-dedication of the work to Akbar.

The Cambridge manuscript contains a paragraph describing the *Ḥamza-nāma* in which occurs the sentence: "In spite of this, in every two years, one volume is prepared." Unless this is a statement uncritically taken from the *Nafāʾis,* or was made at an earlier date (the date of the inception of the manuscript being unknown), it would indicate that work on the *Ḥamza-nāma* was in progress in A. H. 986/ March 1578–9, and that it continued at the same pace as in the days when Mīr Sayyid ʿAlī was in charge of the project. This we know is incorrect, because Mīr ʿAlā al-Daula states that while under Mīr Sayyid ʿAlī work had proceeded at the rate of four volumes in seven years (a little less than two volumes per year), the speed of work had picked up considerably under K͟hwāja ʿAbd al-Ṣamad (see Appendix B).

The Rampur manuscript, according to the editors, is later than the Cambridge text, including events upto the year A. H. 988/ Feb. 1580–1. It was composed sometime after that date, but before A. H. 992/ 1584 as it refers to the famous ʿAbd al-Raḥīm K͟hān b. Bairām K͟hān by the title of Mirzā K͟hān and not by the grand title of K͟hān-i K͟hānān which he received early in A. H. 992/ 1584. In this manuscript, the statement: "In spite of this, in every two years, one volume is prepared," has been crossed out *(qalamzad).* The reasons for this are unclear for the sentence is allowed to stand in the Cambridge manuscript. Perhaps the statement was considered to be erroneous and for that reason cancelled. It is also possible that the statement had become inoperative, the *Ḥamza-nāma,* as we know from other sources, being certainly completed by and not continued beyond A. H. 987.[3]

Muḥammad ʿĀrif's work is not as reliable as that of Mīr ʿAlā al-Daula who was a fine and critical scholar more closely connected to the court and writing of things from first hand knowledge. Muhammad ʿĀrif, on the other hand, seems to have drawn his information second hand. Indeed one of his major sources seems to have been, as noted above, Mīr ʿAlā al-Daula himself. The account of the *Ḥamza-nāma,* however, occurs in a section for which the author claims individuality, and it does contain information which is quite different from that given by Mīr ʿAlā al-Daula. It is interesting to note that the *Ḥamza-nāma* has been chosen by the author to exemplify Akbar's excellence as a "designer of marvels," and is another testimony to

3. *Supra,* p. 66.

the great esteem in which this manuscript was held in the Akbar period itself. Professor C. M. Naim has translated the passage as it occurs on pp. 45–46 of the published text. A translation by Maulānā Imtiyaz Ali Arshi has been reproduced in Ananda Krishna, "A Reassessment," p. 243, fn 15.

"The emperor is a designer of marvels since he has ordered that of the story of Amīr Ḥamza *(qiṣṣa-i amīr ḥamza),* which has 360 tales, each tale should be illustrated with large compositions *(majālis).* Close to one hundred matchless painters *(muṣawwir),* gilders *(muẕahhib),* illuminators *(naqqāsh),* and binders *(mujallid)* are working on that book.[4] The size *(qaṭᶜ)* of that book is one 'meter' and a half *(yak gaz-o-nīm-i sharᶜī);*[5] its paper is imbued with colours; its borders have floral designs *(jul-kārī);*[6] and between two sheets of paper a sheet of *chautār* cloth has been placed to make it more permanent.[7] All the pages are illustrated and gilded. It has been ordered that *munshīs*[8] possessing unique eloquence and a sweet tongue should narrate the entire story in measured and rhythmic prose and that mercury-paced fine calligraphers should put it down in the book. In spite of this, in every two years, one volume is prepared;[9] and on each volume close to one million *tanka-i siyāh* are spent."

4. According to Mīr ᶜAlā al-Daula, the painters numbered thirty (Appendix B). The Ḥājī does not give an exact figure, but even allowing for other craftsmen, the number of painters appear to have increased at the time he was writing.
5. The standard of measurement is different from that used by ᶜAlā al-Daula who gives the size as one *zarᶜ* by one *zarᶜ* (Appendix B).
6. No such borders are to be seen in the paintings as they are preserved now.
7. This is not quite correct. The paintings are on cloth. The text on the reverse has been written on paper which has been pasted on to this cloth.
8. According to Mīr ᶜAlā al-Daula the text was prepared by only one *munshī,* namely Khwāja Aṭāullah (Appendix B).
9. The sentence has been crossed over *(qalamzad)* in the Rampur manuscript.

APPENDIX E
A Mughal Manuscript of the Reign of Humāyūn

In order to get an expert opinion on the Persian features of the *Khamsa* of Niẓāmī discussed in the text [*supra*, pp. 44–46], I consulted my friend Mr. Stuart C. Welch who had also seen the manuscript and with whom I had on occasion talked about its various interesting features. To my great pleasure he offered to write a short note on it with particular reference to two miniatures that he was able to identify as the work of Mīr Muṣawwir, on whose style he has been at work for several years in connection with his researches on the *Shāh-nāma* of Shāh Ṭahmāsp. If his attributions are correct, and I suspect that they are, the manuscript would have to be made for the Mughal court and this would provide convincing evidence for the existence of Indo-Persian work there. This in turn would most satisfactorily account for the presence of several stylistic features derived from it in the Cleveland *Ṭūṭī-nāma.* Of course, the possibility of other Indo-Persian sources contributing to the Mughal style proper as it developed under Akbar would not be excluded for a large number of artists were recruited by the emperor and some of them could well have come from other centers where Indo-Persian painting was patronized. The character of the *Khamsa* manuscript would also serve to emphasise the position of Akbar as the real founder of the Mughal school, for even while it would demonstrate the kind of work prevalent in his father's atelier, it would prove that it was Akbar who was the first to break away from the strong Persian bias, opening up Mughal patronage to all varieties of Indian artistic experience. The Cleveland *Ṭūṭī-nāma* is a precious record of what happened when this was done.

Mr. Welch's note is reproduced below:

"When I first saw the Lalbhai Niẓāmī manuscript fifteen or so years ago, such miniatures as *Majnūn's agony at Lailā's tomb* (Pl. 105) convinced me that the manuscript is Indian rather than Iranian on the basis of such characteristics as the palette, composition, treatment of architecture, and figure painting. At the time, I could not study the volume at length and had little more to say about it. Now having learned far more about both Indian and Iranian painting, I have far more to say.

"*Majnūn's agony* still strikes me as the most appealing and moving of a series of thirteen miniatures in the volume that can be assigned to the same artist or workshop. Although these miniatures fit neatly into the vast tradition of Islamic painting which crosses national borders, they proclaim their Indian-ness in many ways, and bring to mind such other Indian manuscripts of the sixteenth century as the *Candāyana* in the Prince of Wales Museum, the Victoria and Albert Museum *Anwār-i Suhailī,* the *Sindbād-nāma* in the India Office Library, and the *Nujūm al-ᶜUlūm* of the Chester Beatty Library. In these miniatures, the colours are often arranged in fervid combinations that would have offended the more Apollonian Iranian taste. This extra intensity and expressiveness is also seen in the nervously charged silhouettes of flowers, trees, and other vegetation, and in the almost excessively dramatic gestures of the figures, as in Majnūn's skeletal gesticulations. Also markedly Indian is the division of several of these compositions into separate registers, a feature of many Rajasthani and Central Indian styles. Another Indian element is the architecture, with its frequent device of small white domes (Pl. 105), stock items in later Indian work, as in central India and Bundi. An earmark of the particular artist or artists under consideration is the extremely white skin of most of his figures.

"Another group of miniatures in this Niẕāmī manuscript can be assigned to a painter, or painters, trained in the idiom of Bukhara. A throne scene (Pl. 100) is so quintessentially in the mid-sixteenth century mode of the Uzbek capital that were it not in an Indian manuscript, we would assign it to Bukhara. The gold sky, delicate wooden framework over the throne, square pool, doll-like almost round faces, small pointed black boots, palette, and composition are all purely Bukharan. Presumably, the painter[s] were born and trained in Bukhara, where they followed the style brought there by S̱haikẖzāda and other masters from Herat, a somewhat frozen version of the style of the great Bihzād (see Welch, *King's Book of Kings,* p. 60).

"A third group of miniatures is equally un-Indian in character. *A battle scene* (Pl. 98) typifies this strongly Iranian series, although it is somewhat less eclectic than the others and can be linked stylistically to the many battle scenes painted by S̱hāh Ṭahmāsp's lesser artists for his sumptuously illustrated, but qualitatively uneven, *S̱hāh-nāma* (see Dickson and Welch, *The Houghton Shah-nameh,* Harvard University Press, forthcoming).

"Two further miniatures in the Lalbhai Niẕāmī stand in an even closer relationship to the S̱hāh Ṭahmāsp's *S̱hāh-nāma. Naushīrwān listening to the owls on the ruined palace* (Pl. 97) can be attributed to Mīr Muṣawwir, one of the three senior artists—along with Sulṭān Muḥammad and Āqā Mīrak—who worked on the S̱hāh's magnificent version of the Iranian national epic. This artist's style is extremely personal and therefore easily recognisable (for examples of his work see Welch, *op. cit.,* pp. 100–103, 128–30, 168–71; he is discussed at length in Dickson and Welch, *op. cit.).* Essentially a classicist, his characterizations of man and nature are calm, subtle, and extremely harmonious. A careful, patient craftsman, all his autograph miniatures have the same over-all perfection of finish. His pleasing palette often includes an unusual salmon-pink, a particular greenish-blue, and warm yellow—all of which we find in his pictures in the present Niẓāmī, along with his mellifluously elegant line, gracefully rounded arabesques, and even his calligraphic shorthand formula for tufts of grass. Often gently malicious, he amuses us by poking fun at the antics and physiognomies of his subjects, as here in a court scene and in the delightful black mule (Pl. 97).

"Finding Mīr Muṣawwir's work here should not surprise us, other than because of its rarity, for it is known that he had gone to the Mughal court. The Musée Guimet in Paris has a portrait of him by his illustrious son Mīr Sayyid ʿAlī, one of Humāyūn's great Tabriz masters, who joined the Mughal emperor during his exile in Kabul in 1549 and became one of the founders of the Mughal school. This lively portrait depicts Mīr Muṣawwir as a lean, bespectacled, very old graybeard, offering the viewer (his patron, Humāyūn or the boy Akbar presumably) a scroll petitioning better treatment for his son. It is more surprising, perhaps, that Mīr Muṣawwir's Indian period is not otherwise known (except for one other miniature probably from this Niẓāmī manuscript in an Indian private collection). Apparently he was not a major figure in the Mughal ateliers. Abū al-Faẓl, in his section on painters in the *Āʾīn-ī Akbarī,* refers to him only as the father of Mīr Sayyid ʿAlī, and Blochmann's translation has his name as Mīr Manṣūr, presumably a mis-reading of the text. Since he was such an old man at the time of the Guimet portrait, it seems likely that he died or retired soon after it was painted.

"Regrettably, the Lalbhai Niẓāmī offers no precise evidence as to where or when it was copied and illustrated. On the basis of the series of strongly Indian miniatures in what might be termed a Sultanate style (such as our Pl. 105), we can assume that the volume was illustrated in India. Moreover, the inclusion of pictures by Mīr Muṣawwir argues that it was done at the Mughal court, to which the artist is linked by the Guimet portrait. In as much as Humāyūn did not re-enter India until 1555, two years before he died, the manuscript would seem to date to that year or slightly later. The Bukhara and Tabrizi series of illustrations, both in extremely unprogressive modes, are perfectly consistent with such a dating. Moreover, the absence of any pictures in a synthesized style, such as we see in the more progressive illustrations to the Cleveland *Ṭūṭī-nāma,* and the absence of any miniatures derived from non-Muslim Indian sources, supports the argument that this beautiful and important manuscript is the earliest yet known from Mughal India."

SELECT LIST OF ILLUSTRATED MANUSCRIPTS

Akbar-nāma. London, Victoria and Albert Museum. IS 2-1896. 116 miniatures. Ca. 1590.

Amīr Khusrau Dihlawī: ^{c}*Āshiqa* (also known as Dewal Dewī Khiẓr Khān). New Delhi, National Museum. 2 miniatures. Muḥarram 976/ June-July 1568.

Amīr Khusrau Dihlawī: *Khamsa.* Baltimore, Walters Art Gallery. W. 624. 21 miniatures. Dated Year 42 of the Ilāhī era/March 1597–8.

Anwār-i Suhailī. London, British Museum. Add 18 579. 36 miniatures. Ca. 1605–10.

Anwār-i Suhailī. London, School of Oriental Studies. 27 miniatures. Dated 22 Rabī^{c} II 978/ 23 September 1570.

Anwār-i Suhailī. London, Victoria and Albert Museum (ex Erskine of Torrie Institution, Dunimarle Castle, Fife). 126 miniatures. Ca. 1550.

Anwarī: *Dīwān.* Cambridge, Mass, Fogg Art Museum. 15 miniatures. Dated Zu^{c}l-Qa^{c}da 996/ September–October 1588, at Lahore.

Arba^{c}ah of Hātifī. Dublin, Library of A. Chester Beatty. 6 miniatures. Ca. 1550.

Astrology, manuscript on. Rampur, Reza Library. 157 miniatures. Ca. 1570.

Bābar-nāma. New Delhi, National Museum. 145 miniatures. Ca. 1600.

Bhāgavata-purāṇa. Ahmedabad, Private collection. 9 miniatures. Ca. 1575–80.

Bhāgavata-purāṇa. Numerous miniatures dispersed in various collections throughout the world. Ca. 1500–50.

Bhāgavata-purāṇa. Ex collection of the Thakur of Issarda, now dispersed. Ca. 1575 to 1580.

Bhāgavata-purāṇa. Kankroli, Kankroli Maharaj collection. 6 miniatures. Ca. 1590 to 1600.

Candāyana. Banaras, Bharat Kala Bhavan. 6 miniatures. Ca. 1450.

Candāyana. Berlin, Staatsbibliothek. MS Or Fol. 3014. 140 miniatures. Ca. 1440 to 1450.

Candāyana. Lahore and Chandigarh Museums. 24 miniatures. Ca. 1500–50.

Candāyana. Manchester, John Rylands Library. Hindustani MS 1. Over 300 miniatures. Ca. 1500–50.

Candāyana. Bombay, Prince of Wales Museum of Western India. 68 miniatures. Ca. 1500–50.

Caurapañcāśikā. Ahmedabad, N. C. Mehta Gallery of Miniature Paintings. 18 miniatures. Ca. 1500–50.

Dārāb-nāma. London, British Museum. Or 4615. 149 miniatures. Ca. 1580.

Gīta-Govinda. Bombay, Prince of Wales Museum of Western India. No. 54.37–46. 10 miniatures. Ca. 1500–50.

Ḥamza-nāma. See *Qissa-i Amīr Ḥamza.*

ʿIyār-i Dāni<u>sh</u>. Dublin, Library of A. Chester Beatty. Catalogue No. 4. 96 miniatures. Late sixteenth century.

Jāmī: *Bahāristān.* Oxford, Bodleian Library. MS Elliott 254. 6 miniatures. Dated 1595.

Kālacakra-tantra. Cambridge, University Library. Add 1364. Two painted wooden book covers. Dated V. S. 1503/1446, at Arrah, Bihar.

Kālakācārya-kathā. Ahmedabad, Lalbhai Dalpatbhai Institute of Indology. 20 miniatures. Ca. 1435–40.

Kalpasūtra. Bombay, Prince of Wales Museum of Western India. 2 miniatures. Dated V. S. 1423/ 1366, at Delhi.

Kalpasūtra. Baroda, Ātmānanda Jaina Jñāna Mandira. 39 miniatures. V. S. 1522/ 1465, at Jaunpur.

Kalpasūtra. New Delhi, National Museum. 30 miniatures. Dated V. S. 1496/ 1439, at Mandu.

Kalpasūtra and Kālakācārya-kathā. Ahmedabad, Devaśā-no-pāḍā Bhaṇḍāra (Dayāvimalajī Śāstra Saṁgraha). 136 miniatures. Ca. 1475.

Kalpasūtra and Kālakācārya-kathā. Bombay, Prince of Wales Museum of Western India. No. 55. 65. 43 miniatures. Late fourteenth century.

Karaṇḍa-vyūha. Bombay, Haridas Swali Collection. Two painted wooden book covers. Dated V. S. 1512/1455.

Mahābhārata (Āraṇyaka-parva). Bombay, Asiatic Society. Dated 1516, at Kacchauvā.

Mahāpurāṇa. Delhi, Śrī Digambara Nayā Mandira. 65 miniatures. Ca. 1405–25.

Mahāpurāṇa. Jaipur, Śrī Digambara Jaina Atiśaya Kṣetra. 1 miniature. Dated V. S. 1461/ 1404.

Mahāpurāṇa. Jaipur, Śrī Digambara Jaina Atiśaya Kṣetra. Numerous illustrations, 344 folios. Dated 1540, at Palam, near Delhi.

Mahāpurāṇa. Sagar, Jaina Bhaṇḍāra. 290 illustrated folios. Ca. 1510–20.

Miftāḥ al-Fuẓalā. London, British Museum. OR 3299. 187 miniatures. Ca. 1468–75.

Miragāvata of Kutaban. Banaras, Bharat Kala Bhavan. Nos. 7742–7991. 250 miniatures. Ca. 1520–40.

Niʿmat-nāma. London, India Office Library. Pers. MS 149. 50 miniatures. Ca. 1500–1510.

Niẓāmī: *<u>Kh</u>amsa.* Ahmedabad, Kasturbhai Lalbhai Collection. 34 miniatures. ca. 1525–50.

Niẓāmī: *Khamsa.* London, British Museum. Or 2265. 17 miniatures including three additions in the late seventeenth century. Dated 1539–43, made for Shāh Ṭahmāsp.

Niẓāmī: *Khamsa.* London, British Museum, and Baltimore, Walters Art Gallery. 37 miniatures in the British Museum (Or 12 208) and 5 in the Walters Art Gallery (W. 613). Dated Year 40 of the Ilāhī era /1595.

Qiṣṣa-i Amīr Ḥamza. Berlin, Staatsbibliothek. Or 4181. 189 miniatures. Ca. 1450 to 1475.

Qiṣṣa-i Amīr Ḥamza. Vienna, Österreichisches Museum für Angewandte Kunst, 61 paintings; London, Victoria and Albert Museum, 27 paintings; and stray folios in various public and private collections. Probably between 1562 and 1577.

Rāgamālā. Ex Vijayendra Suri collection. 36 miniatures. Ca. 1520–40.

Rāgamālā. Patna, Gopi Krishna Kanoria Collection. Originally consisting of 42 miniatures. Painted at Chawand in 1605 (?).

Razm-nāma. Jaipur, Mahārāja Sawāī Man Singh II Museum. 169 miniatures. Ca. 1582–84.

Saᶜdī: *Būstān.* New Delhi, National Museum. 43 miniatures. Manuscript finished on 15 Rabīᶜ II 908/ 18 October 1502. Name of painter Ḥājī Maḥmūd.

Saᶜdī. *Gulistān.* London, British Museum. Or 5302. 6 miniatures by Shahm Muẕahhib, and 7 miniatures in late Akbar style. Copied at Bukhara in 1566–67.

Saᶜdī. *Gulistān.* London, Royal Asiatic Society. No. 258. 130 folios decorated with birds, animals, and floral designs. Dated 990/ 1581, at Fatehpur-Sikri.

Shāh-nāma. New York, Metropolitan Museum of Art and Arthur C. Houghton. 258 miniatures. Ca. 1520–40, made for Shāh Ṭahmāsp.

Sikandar-nāma. Present whereabouts unknown. Ca. 1450.

Sindbād-nāma. London, India Office Library. Pers MS 3214. 70 miniatures. Ca. 1550.

Tārīkh-i-i Khāndān-i Tīmūriya. Patna, Khuda Baksh Oriental Public Library. No. 551. 132 miniatures. Ca. 1584.

Ṭūṭī-nāma. Banaras, Bharat Kala Bhavan. 9 miniatures. Ca. 1580.

Ṭūṭī-nāma. Cleveland, Ohio, Cleveland Museum of Art. No. 62.279. 211 miniatures. Ca. 1560–65.

Ṭūṭī-nāma. Dublin, Library of A. Chester Beatty. No. 21. 113 illustrations. Ca. 1580.

SELECT BIBLIOGRAPHY

ᶜAbd al-Qādir al-Badāyūnī. *Muntakhab al-Tawārīkh.* Persian text edited by Maulvī Ahmad ᶜAlī. 3 vols. Calcutta, 1868–69.

ᶜAbd al-Qādir al-Badāyūnī. *Muntakhabu-ʾt-Tawārīkh.* English translation by W. H. Lowe, G. Ranking, Wolseley Haig. 3 vols. Calcutta, 1884–1925.

Abū al-Fazl. *Āʾīn-i Akbari by Abūʾl Fazl.* 3 vols. Vol. 1 translated by H. Blochmann, vols 2, 3, by H. S. Jarret, corrected and further annotated by Jadunath Sarkar. Calcutta, 1873, 1949 and 1948.

Abū al-Fazl. *Āʾīn-i Akbari by Abūʾl-Fazl.* Vol. 1. Persian text edited by H. Blochmann. Calcutta, 1877.

Abū al-Fazl. *The Akbar Nāmā of Abu-l-Fazl.* Translated from the Persian by H. Beveridge. 3 vols. Calcutta, 1907–39.

Aḥmad b. Mīr-Munshī. *Calligraphers and Painters: a treatise by Qāḍī Aḥmad,* son of Mīr-Munshī. Translated from the Persian by V. Minorsky. Washington, 1959.

ᶜAlā al-Daula. *Nafāʾis al-Maʾāsir.* Cod. Pers. 3, Bayerische Staatsbibliothek, Munich.

Arberry, A. J., ed. *The Chester Beatty Library: a Catalogue of the Persian Manuscripts and Miniatures.* Vol. 3. Dublin, 1962.

Archer, William G. *Central Indian Painting.* London, 1958.

Archer, William G. and Binney, Edwin 3rd. *Rajput Miniatures from the Collection of Edwin Binney, 3rd.* Portland, Oregon, 1968.

Arnold, Thomas W. *Painting in Islam.* New York, 1965. (Second ed.)

Arnold, Thomas W. and Grohmann, Adolf. *The Islamic Book.* New York, 1929.

Arnold, Thomas W. and Wilkinson, J. V. S. *The Library of A. Chester Beatty: a Catalogue of the Indian Miniatures.* 3. vols. London, 1936.

Aumer, J. *Die Persischen Handschriften der K. Hof- und Staatsbibliothek in München.* Catalogus codicum manu scriptorum Bibliothecae regiae monacensis. Vol. 1, part 3. Munich, 1866.

Bābar, Zahīr al-Dīn Muḥammad. *The Bābur-nāma in English.* Translated by A. S. Beveridge. 2 vols. London, 1922.

Bāyazīd Bīyāt. *Tadhkira-i-Humāyūn wa Akbar of Bāyazīd Biyāt.* A history of the Emperor Humayun from A. H. 949 (A. D. 1542) and of his successor the Emperor Akbar upto A. H. 999 (A. D. 1590). Edited by M. Hidayat Hosain. Calcutta, 1941.

Beveridge, Henry. "The Memoirs of Bayazid (Bajazet) Biyat," *Journal of the Asiatic Society of Bengal* 67 (1898), pp. 296–316.

Binney, Edwin, 3rd. *Indian Miniature Painting From the Collection of Edwin Binney, 3rd: The Mughal and Deccani Schools.* Portland, Oregon, 1973.

Binyon, Laurence. *A Persian Painting of the Sixteenth Century: Emperors and Princes of the House of Timur.* London, 1930.

Binyon, Laurence. *The Poems of Nizami.* London, 1928.

Binyon, Laurence; Wilkinson, J. V. S.; and Gray, Basil. *Persian Miniature Painting including a critical and descriptive catalogue of the miniatures exhibited at Burlington House, January-March, 1931.* London, 1933. (Abbr. BWG).

Boston, Museum of Fine Arts. *Arts of India and Nepal. The Nasli and Alice Heeramaneck Collection.* Boston, 1966.

Brown, Percy. *Indian Painting under the Mughals, A. D. 1550–A. D. 1750.* Oxford, 1924.

Brown, W. Norman, "Some Early Rājasthānī Rāga Paintings," *Journal of the Indian Society of Oriental Art* 16 (1948), pp. 1–10.

Chaghatai, M. Abdulla. "The Illustrated Edition of the Razm Nama," *Bulletin of the Deccan College Research Institute* 5 (1943–4), pp. 281–329.

Chaghatai, M.,Abdulla. "Khawajah ᶜAbd al-Samad Shīrīn-Qalam." *Journal of the Pakistan Historical Society* 11 (1963), pp. 155–181.

Chaghatai, M. Abdulla. "Mir Sayyid Ali Tabrezi." *Pakistan Quarterly* 4 (1954), pp. 24–29 and 60.

Chaghatai, M. Abdulla. *Painting During the Sultanate Period.* Lahore, 1963.

Chaghatai, M. Abdulla. "Tuti Nama (Tales of a Parrot)." *The Pakistan Times,* 20th November 1964, pp. i–ii

Chandra, Moti. "An Illustrated Manuscript of the Kalpasūtra and Kālakācārya-kathā." *Prince of Wales Museum Bulletin* 4 (1953–4), pp. 40–48.

Chandra, Moti. "An Illustrated MS. of Mahapurana in the Collection of Sri Digambar Naya Mandir, Delhi." *Lalit Kala* 5 (April 1959), pp. 68–81.

Chandra, Moti. *Jaina Miniature Paintings from Western India.* Ahmedabad, 1949.

Chandra, Moti. "A Pair of Painted Wooden Covers of the Karandavyuha Ms. Dated A. D. 455 from Eastern India." *Chhavi.* Banaras, 1971, pp. 240–42.

Chandra, Moti. *Studies in Early Indian Painting.* Bombay, 1974

Chandra, Moti and Shah, Umakant P. "New Documents of Jaina Paintings." *Śrī Mahāvīra Jaina Vidyālaya Suvarṇamahotsava Grantha,* Part 1. Bombay, 1968, pp. 348–420.

Chandra, Pramod. "Notes on Mandu Kalpasutra of A. D. 1439." *Marg* 12 (June 1959), pp. 51–54.

Chandra, Pramod. "A Unique Kālakācārya-kathā MS. in the style of the Mandu Kalpasūtra of A. D. 1439." *Bulletin of the American Academy of Benares* 1 (1967), pp. 1–10.

Chandra, Pramod. "Ustad Salivahana and the Development of Popular Mughal Art." *Lalit Kala* 8 (October 1960), pp. 25–46.

Clarke, C. Stanley. *Indian Drawings. Twelve Paintings of the School of Humayun.* London, 1921.

Coomaraswamy, Ananda K. *Miniatures orientales de la collection Goloubew au Museum of Fine Arts de Boston.* Ars Asiatica 13. Paris and Brussels, 1929.

Digby, Simon. "The Bhugola of Ksema Karna: A Dated Sixteenth Century Piece of Indian Metalware." *Art and Archaeology Research Papers* (December 1973), pp. 10–31.

Digby, Simon. "The Literary Evidence for Painting in the Delhi Sultanate." *Bulletin of the American Academy of Benares* 1 (1967), pp. 47–58.

Dimand, W. S. "Several Illustrations from the Dastan-i Amir Hamza in American Collections." *Artibus Asiae* 11 (1948), pp. 5–13.

Doshi, Saryu V. "An Illustrated Ādipurāṇa of 1404 A. D. fromYoginipur." *Chhavi.* Banaras, 1971, pp. 382–391.

Erskine, William. *A History of India Under the Two First Sovereigns of the House of Taimur.* Vol. 2. London, 1854.

Ettinghausen, Richard. "ᶜAbduʾṣ-Ṣamad." *Encyclopaedia of World Art* 1 (1959), pp. 16–20.

Ettinghausen, Richard. "The Bustan Ms. of Sultan Nasīr-Shah Khalji." *Marg* 12 (June 1959), pp. 42–43.

Ettinghausen, Richard. *Paintings of the Sultans and Emperors of India.* New Delhi, 1961.

Fakhruzzaman, Ummahani. "Mir ᶜAlaud-Daulah 'Kami' of Qazvin, his Life and Poetry." *Islamic Culture* 34 (1960), pp. 31–48.

Fraad, Irma L. and Ettinghausen, Richard. "Sultanate Painting in Persian Style, primarily from the first half of the Fifteenth Century. A Preliminary Study." *Chhavi.* Banaras, 1971, pp. 48–66.

Ghani, M. A. *A History of Persian language and literature in the Court of Akbar.* Allahabad, 1929–30.

Glück, Heinrich. *Die Indischen Miniaturen des Haemzae-Romanes im Österreichischen Museum für Kunst und Industrie in Wien und in anderen Sammlungen.* Leipzig, 1925.

Godard, Yedda A. "Les Marges du Murakkaᶜ Gulshan." *Athār-é Īrān* i (1936), pp. 1–33.

Goetz, Hermann. "Vestiges of Muslim Painting under the Sultans of Gujarat." *Journal of the Gujarat Research Society* 16 (July 1954), pp. 212–220.

Gorakshkar, S. V. "A Dated Manuscript of the Kālakācāryakathā in the Prince of Wales Museum." *Prince of Wales Museum Bulletin* 9 (1964–6), pp. 56–57.

Gray, Basil. "The Development of Painting in India in the 16th century." *Marg* 6 (1953), pp. 19–24.

Gray, Basil. "Painting." *Art of India and Pakistan,* a commemorative catalogue of the exhibition held at the Royal Academy of Arts, London, 1947–8. London, 1950, pp. 85–195.

Gray, Basil. *Iran: Persian Miniatures from Ancient Manuscripts.* New York, 1956.

Gray, Basil. "Review of S. C. Welch, The Art of Mughal India." *Artibus Asiae* 28 (1966), pp. 99–101.

Gray, Basil and Barrett, Douglas. *Painting of India.* Geneva, 1963.

Grube, Ernst J. *Islamic Paintings from the 11th to the 18th century in the collection of Hans P. Kraus.* New York, n. d.

Ḥamza-nāma: Vollständige Wiedergabe der Bekannten Blätter der Handschrift aus den Beständen aller Erreichbaren Sammlungen. Vol. 1. Graz, 1974. (Abbr. *Ḥamza*).

Hendley, Thomas H. *Memorials of the Jeypore Exhibition.* Vol. 4. London, 1884.

Husain, Muhammad Ashraf. *A Guide to Fatehpur Sikri.* Delhi, 1937.

Jahāngīr, Nūr al-Dīn Muhammad. *The Tūzuk-i-Jahāngīrī.* Vols. 1 and 2. English translation by A. Rogers and H. Beveridge. London, 1909.

Jauhar. *The Tezkereh al Vakiāt.* English translation by Charles Stewart. London, 1832.

Kanoria, Gopi Krishna. "An Early Dated Rajasthani Rāgamālā." *Journal of the Indian Society of Oriental Art* 19 (1952–3), pp. 1–5.

Khandalavala, Karl. "Leaves from Rajasthan: A Dated Bhagavata Purana of the Bhandarkar Oriental Institute, Poona, and Notes on the Chronology of Early Rajput painting." *Marg* 4 (1950), pp. 2–24 and 49–56.

Khandalavala, Karl. "The Mṛigāvat of Bharat Kala Bhavan." *Chhavi.* Banaras, 1971, pp. 19–36.

Khandalavala, Karl. "Some Problems of Mughal Painting." *Lalit Kala* 11 (April 1962), pp. 9–13.

Khandalavala, Karl and Chandra, Moti. "A Consideration of an Illustrated MS. from Mandapadurga (Mandu) Dated 1439 A.D." *Lalit Kala* 6 (October 1959), pp. 8–29.

Khandalavala, Karl and Chandra, Moti. *An Illustrated Āraṇyaka Parvan in the Asiatic Society of Bombay.* Bombay, 1974.

Khandalavala, Karl and Chandra, Moti. *New Documents of Indian Painting–A Reappraisal.* Bombay, 1969.

Khandalavala, Karl and Chandra, Moti. "Three New Documents of Indian Painting." *Prince of Wales Museum Bulletin* 7 (1959–62), pp. 23–34.

Khandalavala, Karl; Chandra, Moti; and Chandra, Pramod. Miniature Painting: *A Catalogue of the Exhibition of the Sri Motichand Khajanchi Collection held by the Lalit Kala Akademi, 1960.* New Delhi, 1960.

Khandalavala, Karl; Chandra, Moti; Chandra, Pramod; and Gupta P. L. "A New Document of Indian Painting." *Lalit Kala* 10 (1961), pp. 45–54.

Khandalavala, Karl and Mittal, Jagdish. "An Early Akbari Illustrated Manuscript of Tilasm and Zodiac." *Lalit Kala* 14 (1969), pp. 9–20.

Krishna, Anand. "A Reassessment of the Tuti-nama Illustrations in the Cleveland Museum of Art (and Related Problems on Earliest Mughal Paintings and Painters)." *Artibus Asiae* 35 (1973) pp. 241–68.

Krishna, Anand. "Some Pre-Akbari Examples of Rajasthani Illustrations." *Marg* 11 (1958), pp. 18–21.

Krishnadasa, Rai. *Bhārat Kī Chitrakalā.* [In Hindi] Banaras, 1939.

Krishnadasa, Rai. *Muhgal Miniatures.* Delhi, 1955.

Krishnadasa, Rai. "An Illustrated Avadhi MS. of Laur-Chanda in the Bharat Kala Bhavan, Banaras." *Lalit Kala* 1–2 (1955–6), pp. 66–71.

Kühnel, Ernst and Goetz, Hermann. *Indian Book Painting.* London, 1926.

Lee, Sherman E. and Chandra, Pramod. "A Newly Discovered Tuti-Nama and the Continuity of the Indian Tradition of Manuscript Painting." *Burlington Magazine* 105 (December 1963), pp. 547–554.

The Maāthir-ul-Umarā: being biographies of the Muhammadan and Hindu officers of the Timurid sovereigns of India from 1500 to about 1780 A. D. By Shāh Nawāz Khān and his son ᶜAbdul Hayy. Translated by H. Beveridge and revised, annotated and completed by Baini Prashad. 2 vols. Calcutta, 1911–41 and 1952.

Maclagan, Edward. *The Jesuits and the Great Mogul.* London, 1932.

Mahfuz ul-Haq. "Persian Painters, Illuminators and Calligraphists etc. in the XVIth century." *Journal of the Asiatic Society of Bengal* 28 (1932), pp. 239–49.

Martin, F. R. *The Miniature Painting and Painters of Persia, India and Turkey from the 8th to the 18th century.* London, 1912.

Melikian Chirvani, A. S. "L'école de Shiraz et les origines de la miniature moghole." *Paintings from Islamic Lands,* edited by R. Pinder-Wilson. Oxford, 1969, pp. 124 to 41.

Meredith-Owens, G. M. "Ḥamza b. ᶜAbd al-Muṭṭalib." *Encyclopaedia of Islam.* Vol. 3. Leiden, 1971, pp. 132–4.

Muhammad ᶜArif Qandahārī. *Tārikh-i-Akbarī better kown as Tārikh-ī-Qandahārī.* Edited by Muᶜinuʾd-Din Nadwī, Aẓhar Ali Dihlawī and Imtiyāẓ ᶜAli ᶜArshī. Rampur, 1962.

Muḥammad Haidar Dughlat. *A History of the Moghuls of Central Asia being the Tarikh-i-Rashidi of Mirza Muḥammad Haidar, Dughlat.* English version by N. Elias and E. Denison Ross. London, 1898.

Mukherjee, Hiren. "Two Early Rajasthani Rāginī Pictures." *Lalit Kala* 12 (1962), pp. 39–40.

Muqtadir, Abdul. *Catalogue of the Arabic and Persian Manuscripts in the Oriental Public Library at Bankipore.* Vol. 7. Calcutta, 1941.

New Delhi. National Museum of India. *Manuscripts from Indian Collections:* descriptive catalogue. New Delhi, 1964.

Nizami, K. A. "Persian Literature under Akbar." *Medieval India Quarterly* 3 (Nos. 3 and 4, 1958), pp. 300–328.

Nur Bakhsh. "The Agra Fort and its Buildings." *Annual Report of the Archaeological Survey of India,* 1903–4, pp. 164–93.

Pal, Pratapaditya. "A New Document of Indian Painting." *Journal of the Royal Asiatic Society* (1965), pp. 103–111.

Parimoo, Ratan. "A New Set of Early Rajasthani Paintings." *Lalit Kala,* in press.

Pope, A.,U., ed. *Survey of Persian Art.* 6 vols. Oxford, 1939.

Ray, Sukumar. *Humāyūn in Persia.* Calcutta, 1948.

Rieu, Charles. *Catalogue of Persian Manuscripts in the British Museum.* London, 1879–83.

Robinson, B. W. *Descriptive Catalogue of the Persian Painting in the Bodleian Library, Oxford.* Oxford, 1958.

Robinson, B. W. *Persian Miniature Painting from Collections in the British Isles.* London, 1967.

Robinson, B. W. *Persian Paintings.* London, 1952.

Sādiqī Beg Afshār. *Tazkira Majmaʿ al-Khwās.* Tabriz, A. H. 1327.

Sakisian, Armenag Bey. *La miniature Persane du XIIe au XVIIe siècle.* Paris and Brussels, 1929.

Sakisian, Armenag Bey. "The School of Bihzad and the Miniaturists Aqa Mirak and Mir Musavvir." *Burlington Magazine* 68 (February, 1936), pp. 80–85.

Schroeder, Eric. *Persian Miniatures in the Fogg Museum of Art.* Cambridge, Mass., 1942.

Shiveshwarkar, Leela. *The Pictures of the Chaurapañcāśikā:* A Sanskrit love lyric. New Delhi, 1967.

Skelton, Robert. "The Mughal Artist Farrokh Beg." *Ars Orientalis* 2 (1957), pp. 393–411.

Skelton, Robert. "Mughal Paintings from the Harivamsa Manuscript." *Victoria and Albert Museum Year Book* 2, London, 1970, pp. 41–54.

Skelton, Robert. "The Niʿmat nama: A Landmark in Malwa Painting." *Marg* 12 (June 1959), pp. 44–50.

Smith, Edmund W. *The Moghul Architecture of Fatehpur Sikri.* 4 vols. Allahabad, 1897.

Smith, Vincent. *Akbar, the Great Mogul 1542–1605.* 2nd edition, reprint. Delhi, 1958.

Spink, Walter M. *Krishnamandala:* A devotional theme in Indian Art. Ann Arbor, Mich., 1971.

Sprenger, Aloys. *A Catalogue of the Arabic, Persian and Hindustany Manuscripts of the Libraries of the King of Oudh.* Vol. 1 containing Persian and Hindustany poetry. Calcutta, 1854.

Srivastava, A. L. *Akbar the Great.* 2 vols. Agra, 1962–7.

Staude, W. "Les Artistes de la Cour d'Akbar et les illustrations du Dastan i-Amir Hamzah." *Arts Asiatiques* 2 (1955), pp. 47–65.

Staude, W. "Contribution à l'etude de Basawan." *Revue des Arts Asiatiques* 8 (1934), pp. 1–18.

Stchoukine, Ivan. *Miniatures Indiennes du Musée du Louvre.* Paris, 1929.

Stchoukine, Ivan. *La Peinture Indienne à l'époque des grands Moghols.* Paris, 1929.

Stchoukine, Ivan. *Les peintures des manuscrits Safavis de 1502 à 1587.* Paris, 1959.

Stchoukine, Ivan; Flemming, Barbara; Luft, Paul; and Sahrweide, Hanna. *Illuminierte Islamische Handschriften.* Wiesbaden, 1971.

Storey, C. A. *Persian Literature:* A Bio-Bibliographical Survey. London, 1953.

Tarafdar, M. R. "Illustrations of the Chandain in the Central Museum, Lahore." *Journal of the Asiatic Society of Pakistan* 8 (December 1963), pp. 109–115.

Taraporevala, V. D. B. and Marshall, D. N. *Mughal Bibliography.* Bombay, 1962.

Titley, Norah M. "An Illustrated Persian Glossary of the Sixteenth Century." *British Museum Quarterly* 29 (1965), pp. 15–19.

Welch, Stuart C. *The Art of Mughal India.* New York, 1963.

Welch, Stuart C. "Early Mughal Miniature Paintings from Two Private Collections shown at the Fogg Art Museum." *Ars Orientalis* 3 (1959), pp. 133–146.

Welch, Stuart C. "The Emperor Akbar's Khamsa of Nizami." *Journal of the Walters Art Gallery* 23 (1960), pp. 87–96.

Welch, Stuart C. *A Flower From Every Meadow:* Indian Paintings from American Collections. New York, 1973.

Welch, Stuart C. *A King's Book of Kings:* The Shah-namah of Shah Tahmasp. New York, 1972.

Welch, Stuart C. "The Paintings of Basawan." *Lalit Kala* 10 (October 1961), pp. 7–17.

Welch, Stuart C. and Beach, Milo C. *Gods, Thrones and Peacocks.* New York, 1965.

Wilkinson, J. V. S. *The Lights of Canopus.* London, n. d.

Wilkinson, J. V. S. and Gray, Basil. "Indian Paintings in a Persian Museum." *Burlington Magazine* 66 (1935), pp. 168-177.

LIST OF PLATES

References in parentheses are to illustrations of the full miniatures published in Glück and *Ḥamza-nāma;* see bibliography.

Plate

1.	*Ḥamza-nāma,* detail.	Boston Museum of Fine Arts.
2.	*Ḥamza-nāma,* detail.	Brooklyn Museum.
3.	*Ḥamza-nāma,* detail.	Freer Gallery of Art, Washington D. C.
4.	*Ḥamza-nāma,* detail.	Los Angeles County Museum.
5.	*Ḥamza-nāma,* detail.	Los Angeles County Museum.
6.	*Ḥamza-nāma,* detail.	Los Angeles County Museum.
7.	*Ḥamza-nāma,* detail.	Metropolitan Museum of Art, New York.
8.	*Ḥamza-nāma,* detail.	Free Library, Philadelphia.
9.	*Ḥamza-nāma,* detail.	Victoria and Albert Museum, London. No. IS 7-1949.
10.	*Ḥamza-nāma,* detail.	Victoria and Albert Museum, London. No. IS 7-1949.
11.	*Ḥamza-nāma,* detail.	Victoria and Albert Museum, London. (Glück, Abb. 1).
12.	*Ḥamza-nāma,* detail.	Victoria and Albert Museum, London. (Glück, Abb. 3).
13.	*Ḥamza-nāma,* detail.	Victoria and Albert Museum, London. (Glück, Abb. 8).
14.	*Ḥamza-nāma,* detail.	Victoria and Albert Museum, London. (Glück, Abb. 15).
15.	*Ḥamza-nāma,* detail.	Museum für Angewandte Kunst, Vienna. (Glück, Abb. 20; *Ḥamza,* Pl. 12).
16.	*Ḥamza-nāma,* detail.	Victoria and Albert Museum, London. (Glück, Abb. 33).
17.	*Ḥamza-nāma,* detail.	Victoria and Albert Museum, London. (Glück, Abb. 40).
18.	*Ḥamza-nāma,* detail.	Victoria and Albert Museum, London. (Glück, Abb. 43).
19.	*Ḥamza-nāma,* detail.	Museum für Angewandte Kunst, Vienna. (Glück, Taf. 3; *Ḥamza;* Pl. 5).
20.	*Ḥamza-nāma,* detail.	Museum für Angewandte Kunst, Vienna. (Glück, Taf. 7; *Ḥamza;* Pl. 13).
21.	*Ḥamza-nāma,* detail.	Museum für Angewandte Kunst, Vienna. (Glück, Taf. 11; *Ḥamza,* Pl. 17).
22.	*Ḥamza-nāma,* detail.	Museum für Angewandte Kunst, Vienna. (Glück, Taf. 12; *Ḥamza,* Pl. 18).
23.	*Ḥamza-nāma,* detail.	Museum für Angewandte Kunst, Vienna. (Glück, Taf. 13; *Ḥamza,* Pl. 19).
24.	*Hamza-nāma,* detail.	Museum für Angewandte Kunst, Vienna. (Glück, Taf. 15; *Ḥamza,* Pl. 21).
25.	*Ḥamza-nāma,* detail.	Museum für Angewandte Kunst, Vienna. (Glück, Taf. 17; *Ḥamza,* Pl. 23).
26.	*Ḥamza-nāma,* detail.	Museum für Angewandte Kunst, Vienna. (Glück, Taf. 17; *Ḥamza,* Pl. 23).
27.	*Ḥamza-nāma,* detail.	Museum für Angewandte Kunst, Vienna. (Glück, Taf. 19; *Ḥamza,* Pl. 25).
28.	*Ḥamza-nāma,* detail.	Museum für Angewandte Kunst, Vienna. (Glück, Taf. 21; *Ḥamza,* Pl. 27).
29.	*Ḥamza-nāma,* detail.	Museum für Angewandte Kunst, Vienna. (Glück, Taf. 24; *Ḥamza,* Pl. 30).
30.	*Ḥamza-nāma,* detail.	Museum für Angewandte Kunst, Vienna. (Glück, Taf. 26; *Ḥamza,* Pl. 32).
31.	*Ḥamza-nāma,* detail.	Museum für Angewandte Kunst, Vienna. (Glück, Taf. 33; *Ḥamza,* Pl. 40).
32.	*Ḥamza-nāma,* detail.	Museum für Angewandte Kunst, Vienna. (Glück, Taf. 34; *Ḥamza,* Pl. 41).
33.	*Ḥamza-nāma,* detail.	Museum für Angewandte Kunst, Vienna. (Glück, Taf. 37; *Ḥamza,* Pl. 44).
34.	*Ḥamza-nāma,* detail.	Museum für Angewandte Kunst, Vienna. (Glück, Taf. 40; *Ḥamza,* Pl. 47).
35.	*ʿĀs̲h̲iqa,* dated 1568.	National Museum, New Delhi.

36. *ʿĀshiqa*, dated 1568. National Museum, New Delhi.
37. *Gulistān*, copied 1566-7. Miniature by Shaḥm Muẕahhib. British Museum, London.
38. *Gulistān*, copied 1566-7. Detail of miniature by Shaḥm Muẕahhib. British Museum, London.
39. *Anwār-i Suhailī*, dated 1570. School of Oriental Studies, London.
40. *Anwār-i Suhailī*, dated 1570. School of Oriental Studies, London.
41. *Anwār-i Suhailī*, dated 1570. Detail. School of Oriental Studies, London.
42. *Anwār-i Suhailī*, detail, dated 1570. Detail. School of Oriental Studies, London.
43. *Astrology* MS. Raza Library, Rampur.
44. *Astrology* MS. Raza Library, Rampur.
45. *Dārāb-nāma.* Miniature by Bihzād, refined by ʿAbd al-Ṣamad. British Museum, London.
46. *Dārāb-nāma.* Miniature by Ibrāhīm Lāhorī. British Museum, London.
47. *Ṭūṭī-nāma.* Library of A. Chester Beatty, Dublin.
48. *Ṭūṭī-nāma.* Library of A. Chester Beatty, Dublin.
49. *Ṭūṭī-nāma.* Library of A. Chester Beatty, Dublin.
50. *Ṭūṭī-nāma.* Library of A. Chester Beatty, Dublin.
51. *Ṭūṭī-nāma.* Library of A. Chester Beatty, Dublin.
52. *Ṭūṭī-nāma.* Library of A. Chester Beatty, Dublin.
53. *Ṭūṭī-nāma.* Library of A. Chester Beatty, Dublin.
54. *Ṭūṭī-nāma.* Library of A. Chester Beatty, Dublin.
55. *Ṭūṭī-nāma.* Library of A. Chester Beatty, Dublin.
56. *Ṭūṭī-nāma.* Library of A. Chester Beatty, Dublin.
57. *Ṭūṭī-nāma.* Library of A. Chester Beatty, Dublin.
58. *Ṭūṭī-nāma.* Library of A. Chester Beatty, Dublin.
59. *Ṭūṭī-nāma.* Library of A. Chester Beatty, Dublin.
60. *Ṭūṭī-nāma.* National Museum, New Delhi.
61. *Ṭūṭī-nāma.* Library of A. Chester Beatty, Dublin.
62. *Akbar-nāma.* Detail. Victoria and Albert Museum, London.
63. *Akbar-nāma.* Detail. Victoria and Albert Museum, London.
64. *Portrait of a young Indian scholar.* Collection of Edwin Binney 3rd.
65. *Jamshīd writing on a rock.* Miniature by Khwāja ʿAbd al-Ṣamad, dated 1588. Freer Gallery of Art, Washington.D. C.
66. *Princes of the House of Tīmūr.* Detail. British Museum, London.
67. *Kālakācārya-kathā.* L. D. Institute of Indololgy, Ahmedabad.
68. *Kālakācārya-kathā.* L. D. Institute of Indololgy, Ahmedabad.
69. *Candāyana.* Berlin Staatsbibliothek.
70. *Candāyana.* Berlin Staatsbibliothek.
71. *Ḥamza-nāma.* Berlin Staatsbibliothek.
72. *Ḥamza-nāma.* Berlin Staatsbibliothek.
73. *Mahābhārata*, dated 1516. Asiatic Society, Bombay.
74. *Mahābhārata*, dated 1516. Asiatic Society, Bombay.
75. *Mahāpurāṇa*, dated 1540. Śrī Digambara Jaina Atiśaya Kṣetra, Jaipur.
76. *Bhāgavata-purāṇa.* Khajanchi collection, Bikaner.
77. *Bhāgavata-purāṇa.* Art Institute of Chicago.
78. *Bhāgavata-purāṇa.* Bharat Kala Bhavan, Banaras.
79. *Bhāgavata-purāṇa.* Cleveland Museum of Art.
80. *Gīta-Govinda.* Prince of Wales Museum of Western India, Bombay.

81. *Caurapañcāśikā.* N. C. Mehta Gallery of Miniature Painting, Ahmedabad.
82. *Caurapañcāśikā.* N. C. Mehta Gallery of Miniature Painting, Ahmedabad.
83. *Caurapañcāśikā.* N. C. Mehta Gallery of Miniature Painting, Ahmedabad.
84. *Caurapañcāśikā.* N. C. Mehta Gallery of Miniature Painting, Ahmedabad.
85. *Bhāgavata-purāṇa.* Collection of Edwin Binney 3rd.
86. *Bhāgavata-purāṇa.* Private collection.
87. *Bhāgavata-purāṇa.* Private collection, Ahmedabad.
88. *Bhāgavata-purāṇa.* Private collection, Ahmedabad.
89. *Bhāgavata-purāṇa.* Private collection, Ahmedabad.
90. *Bhāgavata-purāṇa.* Private collection, Ahmedabad.
91. *Rāga Vasanta.* Bharat Kala Bhavan, Banaras.
92. *Bhāgavata-purāṇa.* Kankroli Maharaj collection.
93. *Bhāgavata-purāṇa.* Kankroli Maharaj collection.
94. *Rāga Gauḍī,* from Chawand Rāgamālā. Kanoria collection, Patna.
95. *Niᶜmat-nāma.* India Office Library, London.
96. *Niᶜmat-nāma.* Detail. India Office Library, London.
97. *Khamsa* of Niẓāmī. Kasturbhai Lalbhai collection, Ahmedabad.
98. *Khamsa* of Niẓāmī. Kasturbhai Lalbhai collection, Ahmedabad.
99. *Khamsa* of Niẓāmī. Kasturbhai Lalbhai collection, Ahmedabad.
100. *Khamsa* of Niẓāmī. Kasturbhai Lalbhai collection, Ahmedabad.
101. *Khamsa* of Niẓāmī. Kasturbhai Lalbhai collection, Ahmedabad.
102. *Khamsa* of Niẓāmī. Kasturbhai Lalbhai collection, Ahmedabad.
103. *Khamsa* of Niẓāmī. Kasturbhai Lalbhai collection, Ahmedabad.
104. *Khamsa* of Niẓāmī. Kasturbhai Lalbhai collection, Ahmedabad.
105. *Khamsa* of Niẓāmī. Kasturbhai Lalbhai collection, Ahmedabad.
106. *Candāyana.* Prince of Wales Museum of Western India, Bombay.
107. *Candāyana.* Prince of Wales Museum of Western India, Bombay.
108. *Candāyana.* Prince of Wales Museum of Western India, Bombay.
109. *Candāyana.* Prince of Wales Museum of Western India, Bombay.
110. *Candāyana.* Prince of Wales Museum of Western India, Bombay.
111. *Candāyana.* Prince of Wales Museum of Western India, Bombay.
112. *Candāyana.* John Rylands Library, Manchester.
113. *Ṭūṭī-nāma.* Bharat Kala Bhavan, Banaras.
114. *Ṭūṭī-nāma.* Bharat Kala Bhavan, Banaras.
115. *Ṭūṭī-nāma.* Infra-red photograph of fol. 115v. Cleveland Museum of Art.
116. *Ṭūṭī-nāma.* Infra-red photograph of fol. 115. Cleveland Museum of Art.
117. *Nafāʾis al-Maʾāṣir.* Fol. 54. Bayerische Staatsbibliothek, Munich.
118. *Nafāʾis al-Maʾāṣir.* Fol. 54v. Bayerische Staatsbibliothek, Munich.
119. *Nafāʾis al-Maʾāṣir.* Fol. 55. Bayerische Staatsbibliothek, Munich.
120. *Nafāʾis al-Maʾāṣir.* Fol. 335. Bayerische Staatsbibliothek, Munich.

List of Tables

Photographic credits

All photographs except those listed below are by the author. Grateful acknowledgements are made to the various persons and institutions for their courtesy in providing photographs and permitting their reproduction.

Mr. Edwin Binney 3rd	Pls. 64, 85
Freer Gallery of Art, Washington, D. C.	65
Staatsbibliothek der Stiftung Preussischer Kulturbesitz, Berlin	69–70
Cleveland Museum of Art	79, 115–116
Dr. Ratan Parimoo	87–90
Bayerische Staatsbibliothek, Munich	117–120

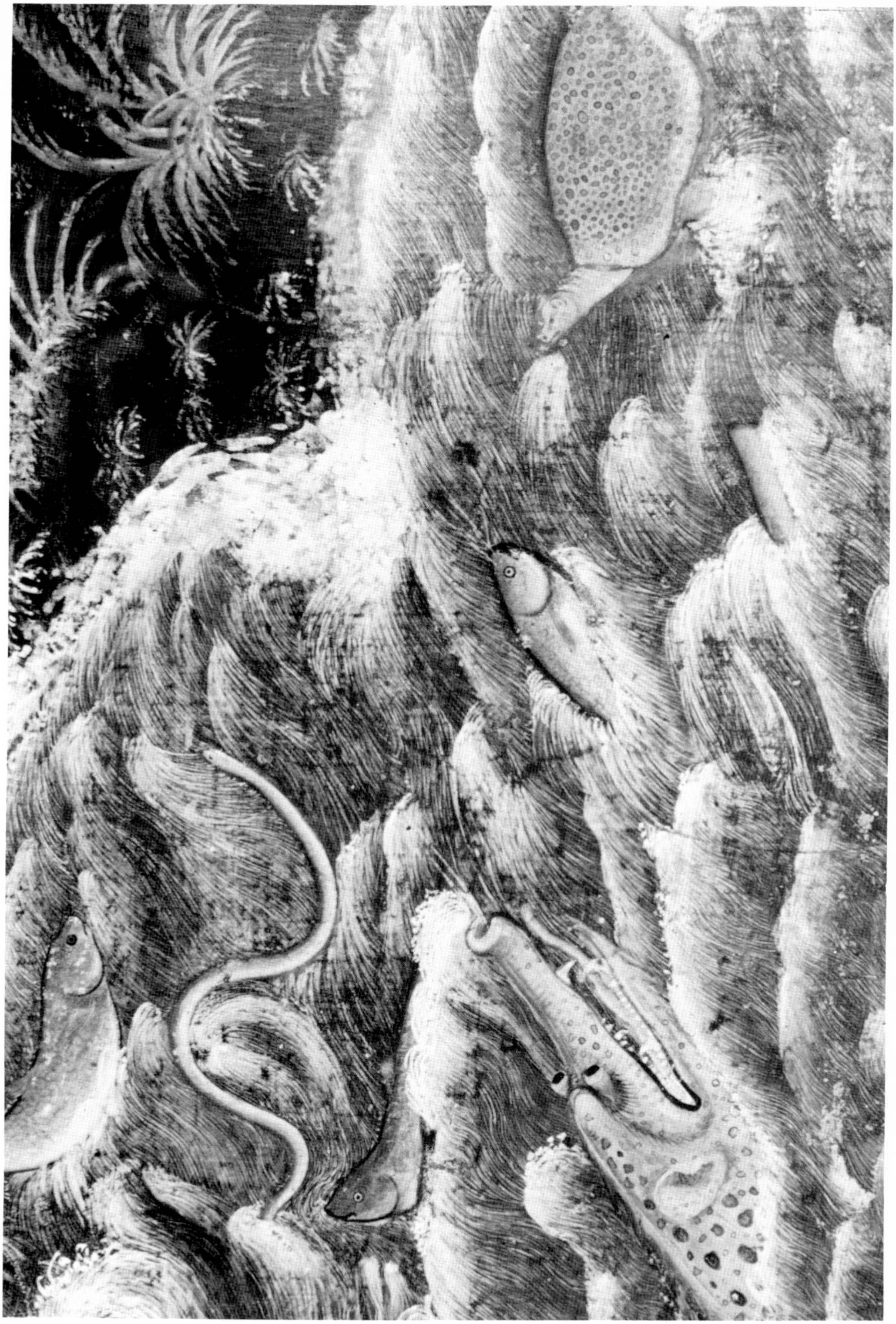

Pl. 2

Pl. 4

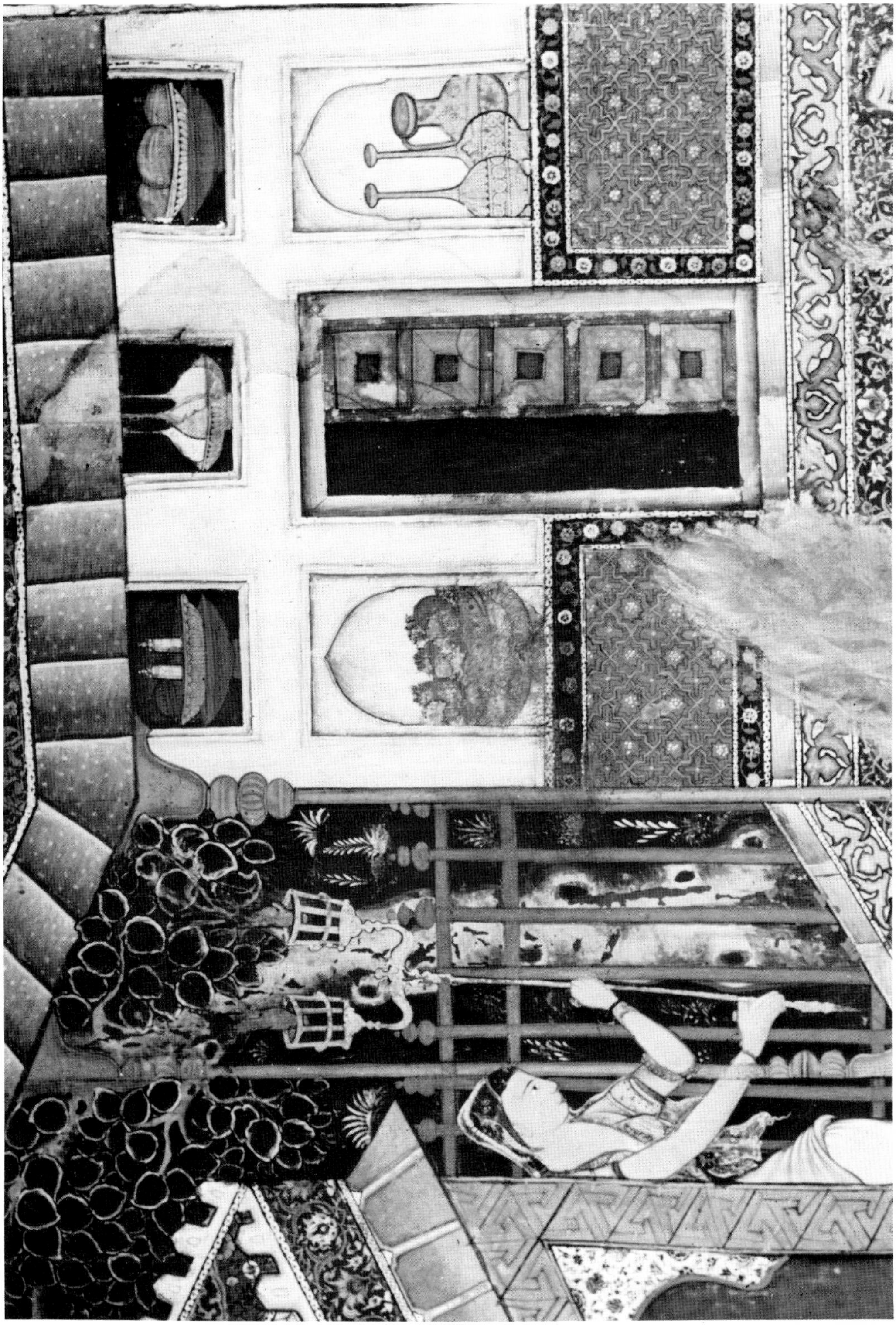

Pl. 6

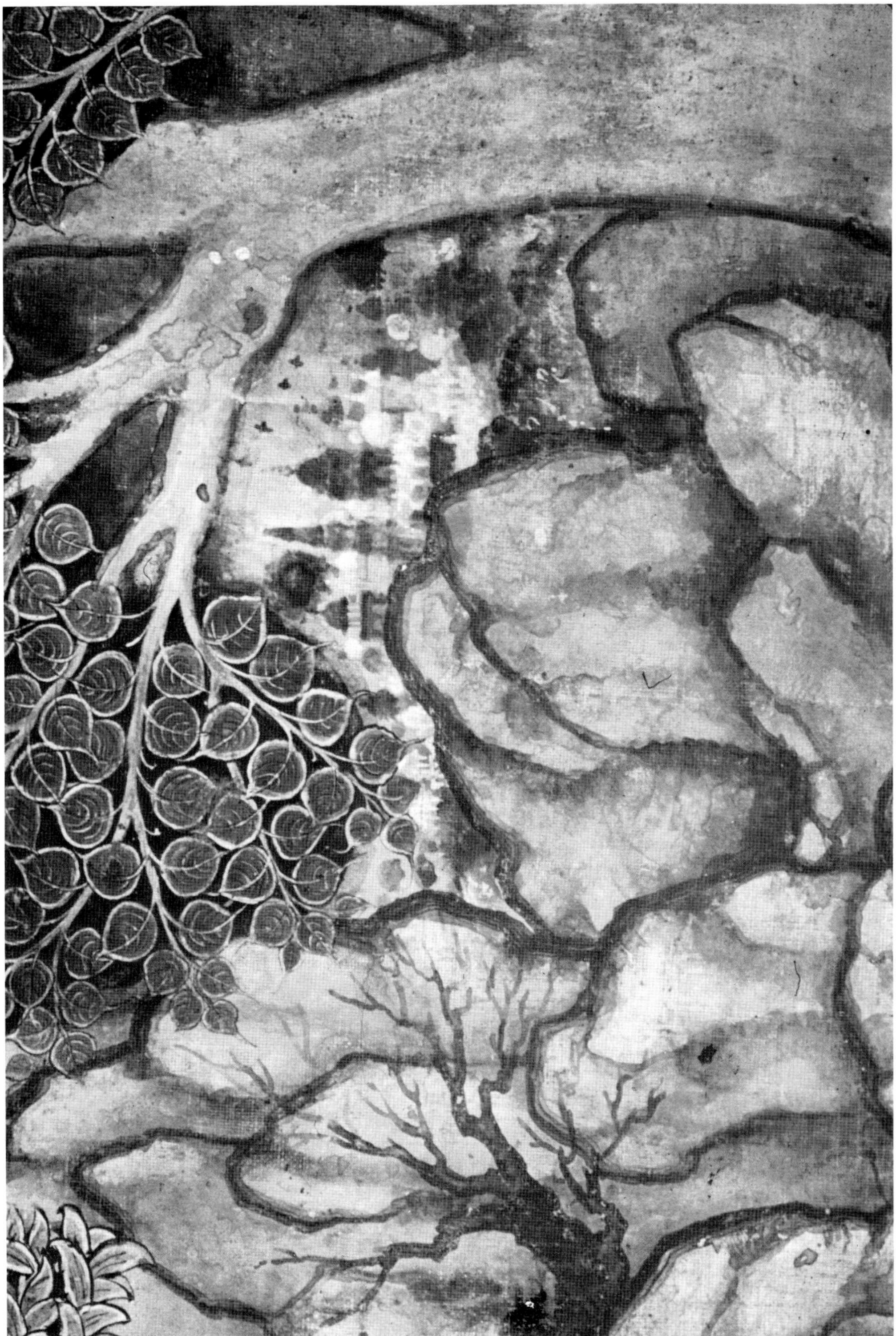

Pl. 10

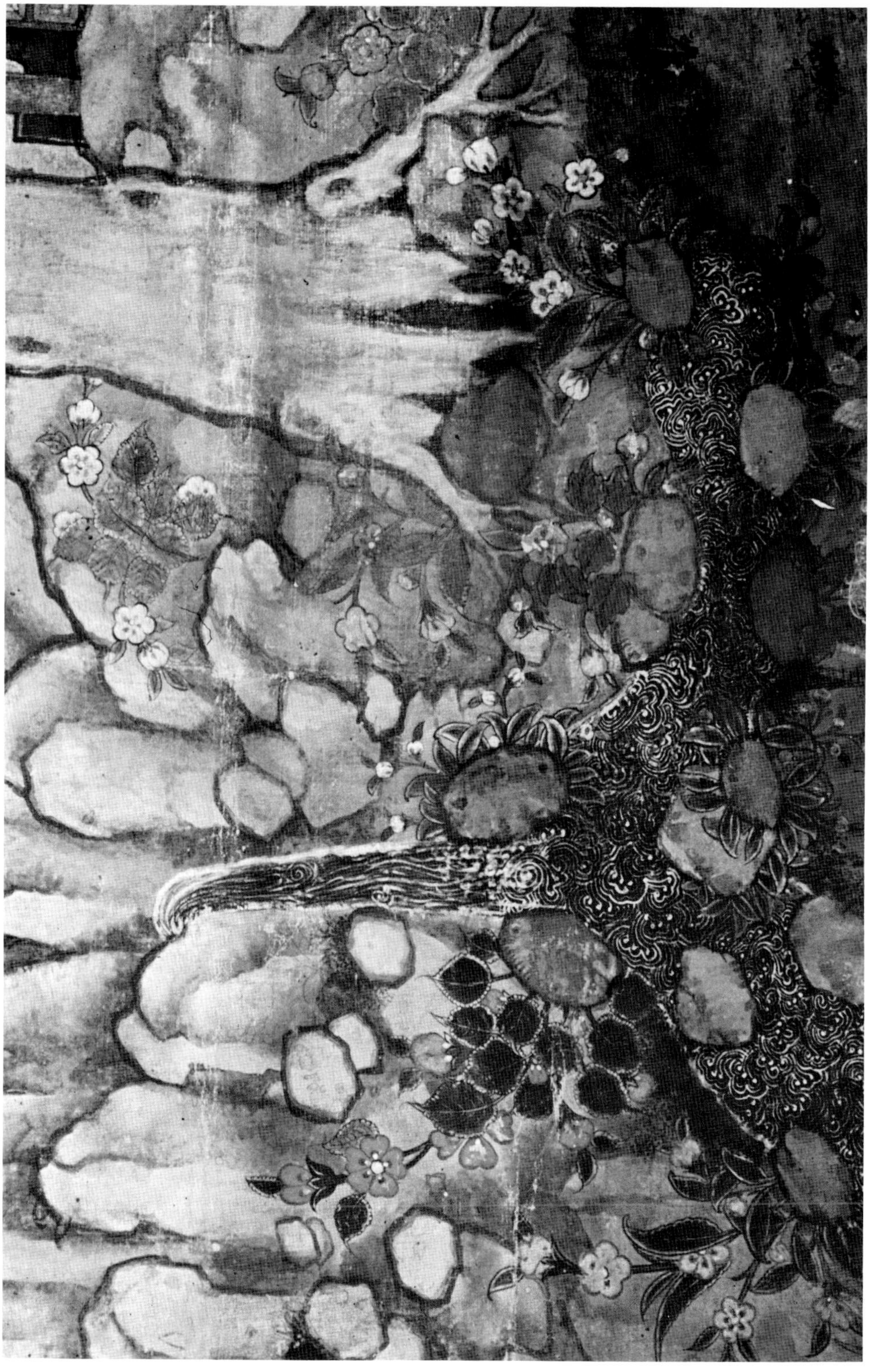

Pl. 18

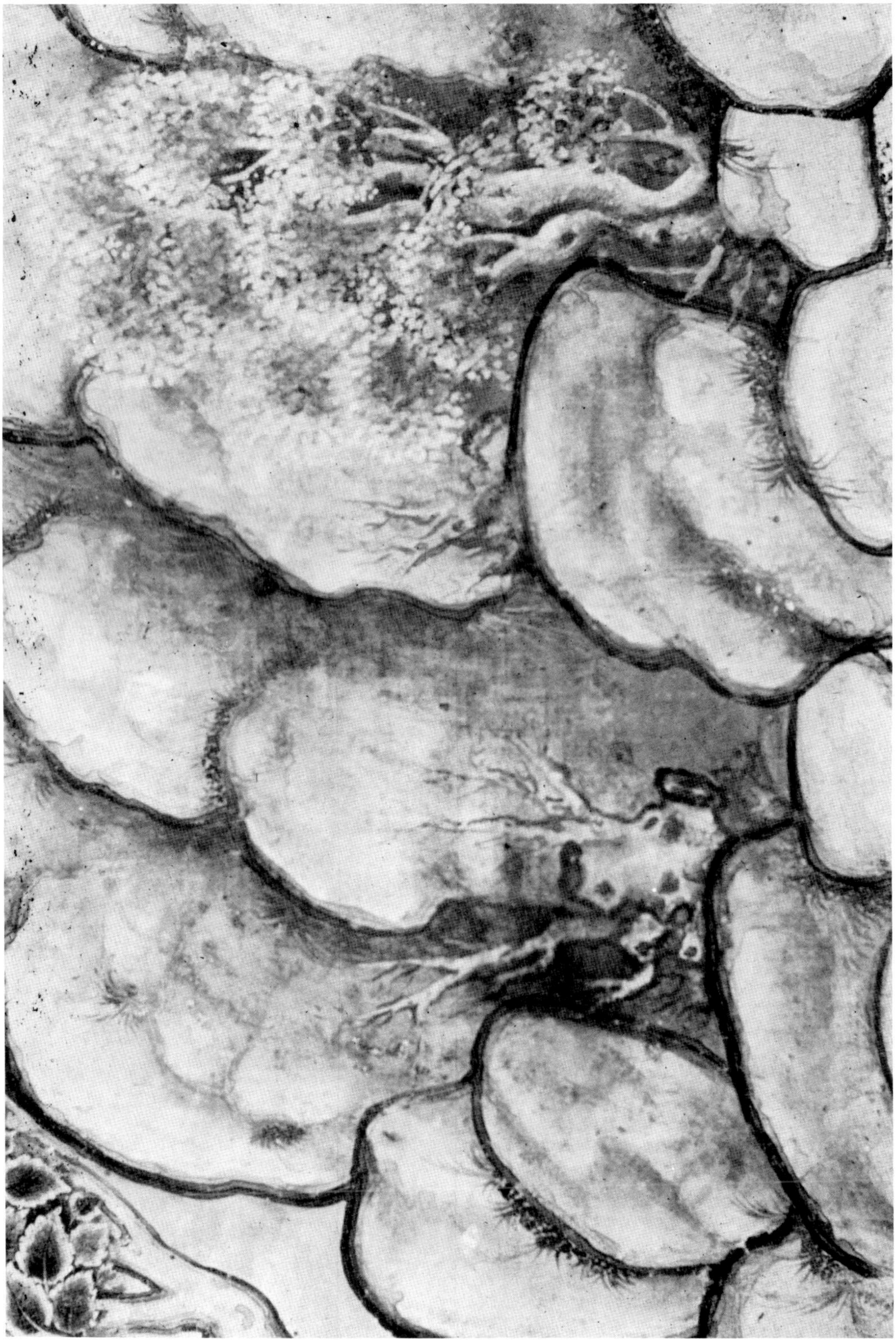

Pl. 26

Pl. 28

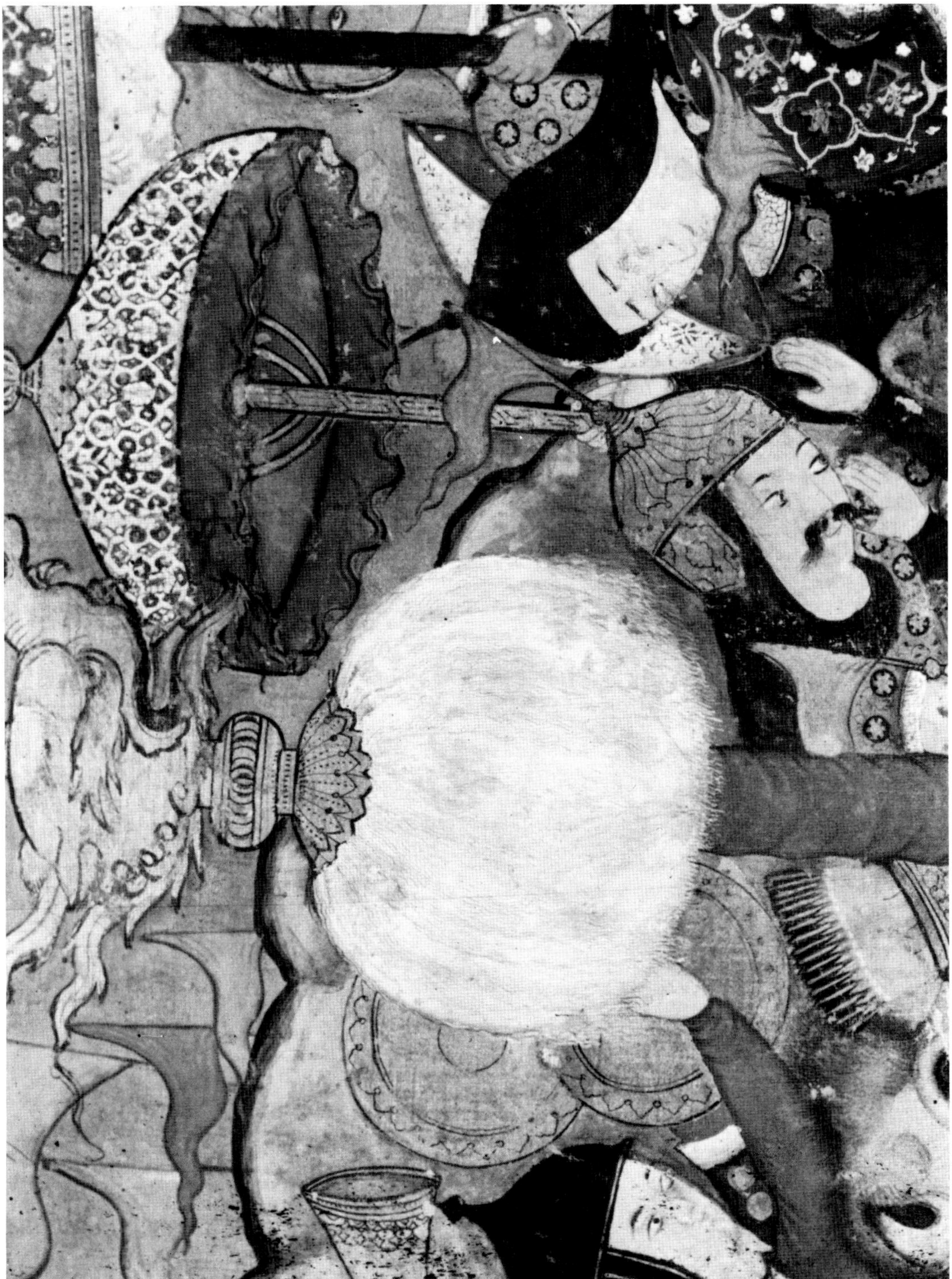

Pl. 36

قاضی دریافت که چه حالتست گفت از کدام جانب برآمد گفت از طرف مشرق گفت الحمد لله که هنوز در توبه باز است بگم این حدیث که

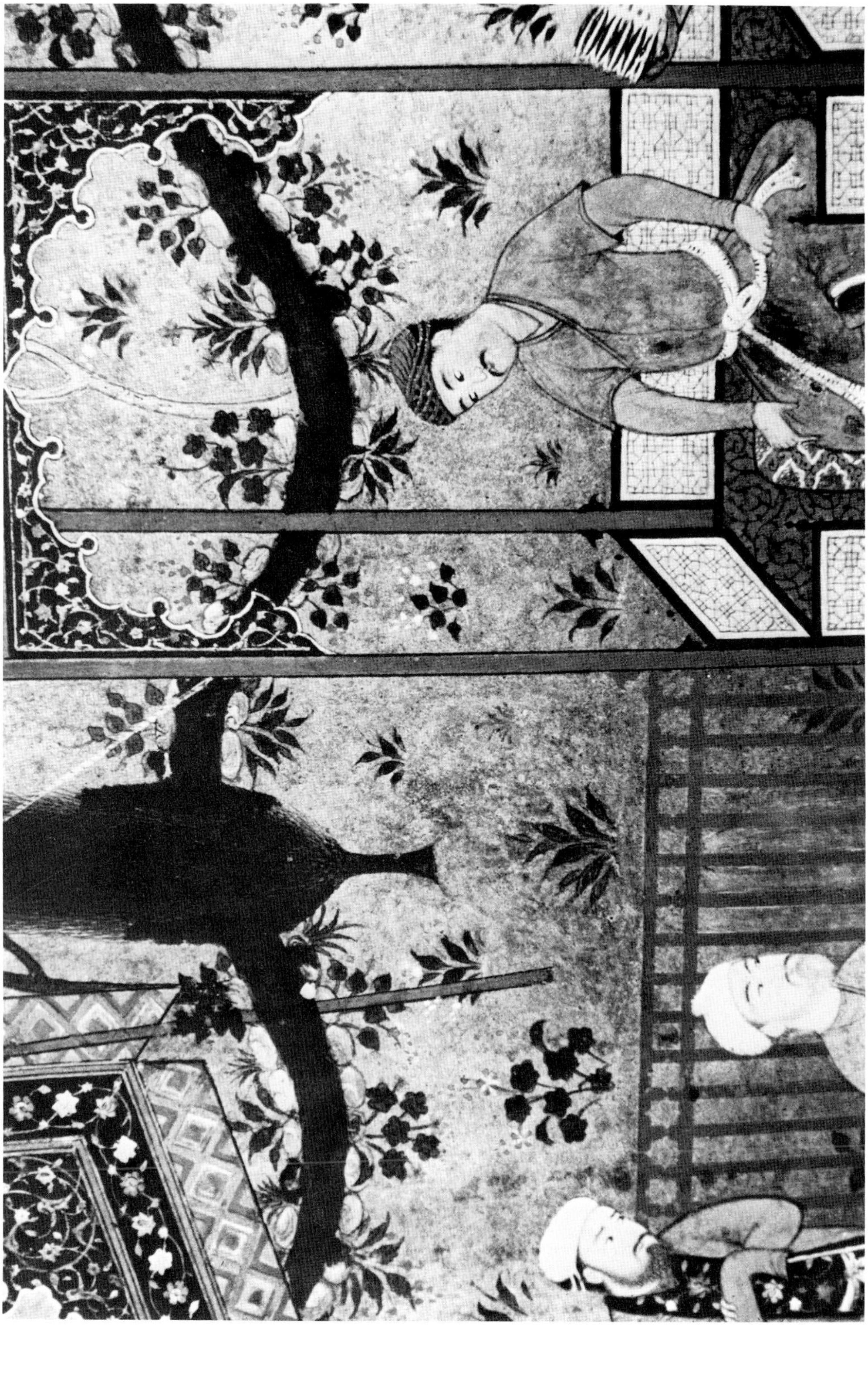

Pl. 40

مقرر شد که زن ازان مبراست در زمانی
او در گذشت و فرمود
که بازدار را بیارید بازدار بازی بر دست گرفت شعفی تمام در آمد که مگر تشریفی
خواهد یافت زن پرسید که ای ستمکار غدار تو دیدی که من کاری خلاف رضای
خدای تعالی میکردم گفت آری من دیدم همین که این کلمه بر زبان راند
بازی که بر دست داشت قصد روی او کرده منقار در چشمش زد و برکند

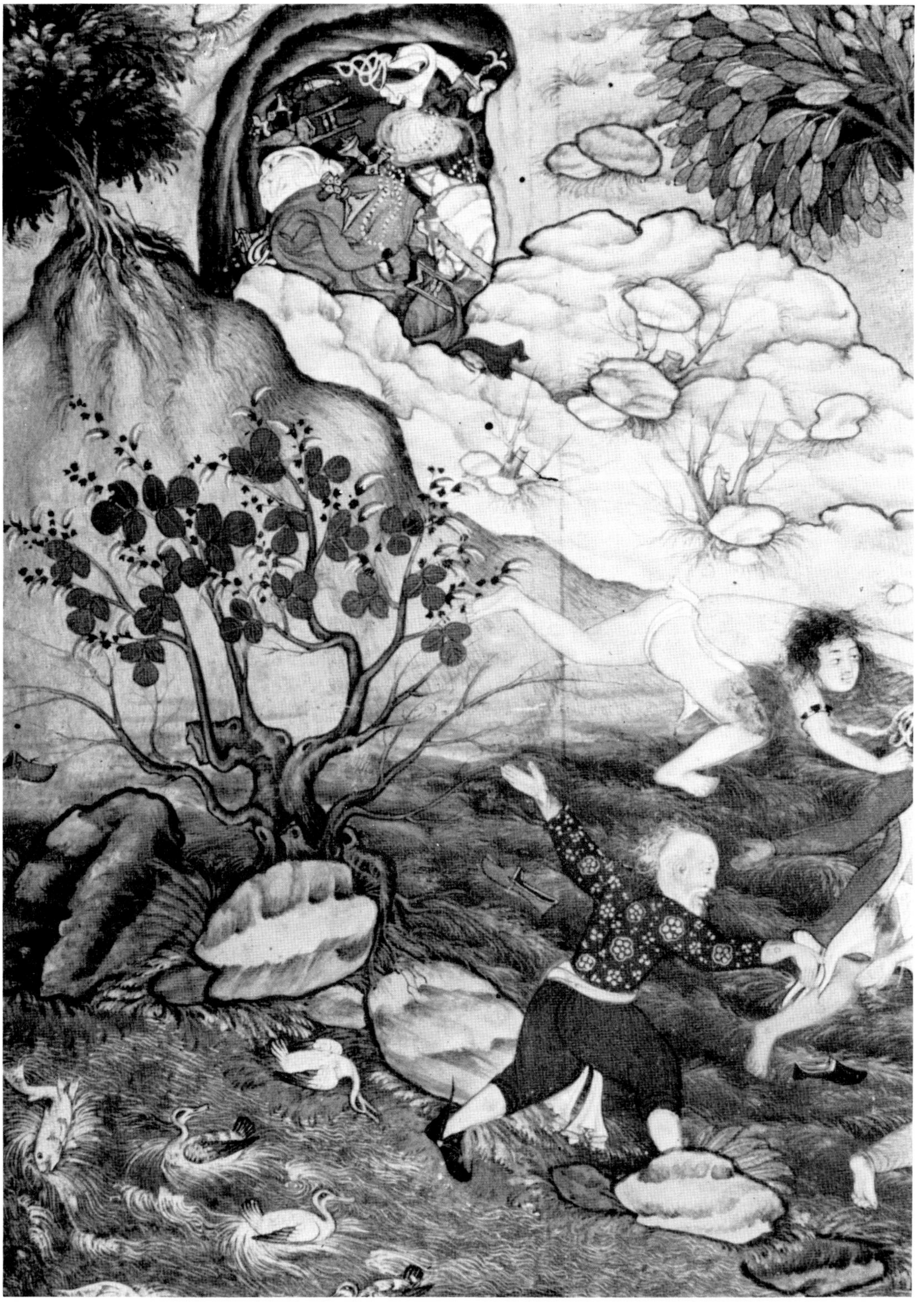

Pl. 42

Pl. 44

ای ملک مارا چنین گفتند که ترا بند کرده‌اند و نزدیک قیصر برده‌اند تو اینجا چه میکنی و این مرد کیست اینجا قصه تمامی
گفت مترس که گماشته نزد دانست که دوش بیامد و مرا از دشت کوه آسا بستد و بدین کوه آورد و رشنواد
گفت ای ملک مرا دوش طرفه کاری افتاده بود بطلایه رفته بودم تمامی گفت بگوی گفت چون بمیان راه
برسیدم برابر طاق مهراب دران وقت که باران می آمد و آوازی شنودم که ای طاق کهنه فرو شو
ساعتی قرار گیر که در بایان تو داراب بسر ارده شیر خفته است و جهان را بمهر او نیازست من عجب داشتم
و بیامدم جوانی را دیدم در زیر طاق خفته و اسبی بر زبر سر او ایستاده آن جوان را بیدار کردم و گفتم
بیرون آی تا بو بینم که تو کیستی آن جوان برخاست و بیرون آمد در ساعت آن طاق فرود آمد و بیفتاد
چون من آن چنان بدیدم عجب داشتم آن جوانمرد را خواستم که نزدیک تو آرم تا بنگری که او
کیست ترا گرفته بودند و بند کرده چون بدین شهر رسیدم من بگریختم و آن جوان را بگذاشتم باید که اکنون
مرا ملک بگوید که احوال و چگونه بوده است و ارده شیر را بسر کجا بوده است تمامی چون این سخن بشنید
گفت ای رشنواد راست میگویی آن فرزند منست آن وقت که او از من جدا شد از بهر تاج و تخت
و پادشاهی او را در تابوتی کردم و بدریا فرو گذاشتم گازری وی را بگرفت و بزرگ کرد پس بدست
مردو افتاد و باز نزدیک من آمد نیارستم گفتن که این فرزند منست که ترسیدم که مرا استوار ندارند و او را
از بر خود بدر رود کردم چند گاه رفته بود و پادشاه یونان و ملکوت شده بود و اکنون باز آمد کار خود شوریده دست

عمل بهزاد اصلاح خواجه عبد الصمد

١٦٨

27

68

٥٨

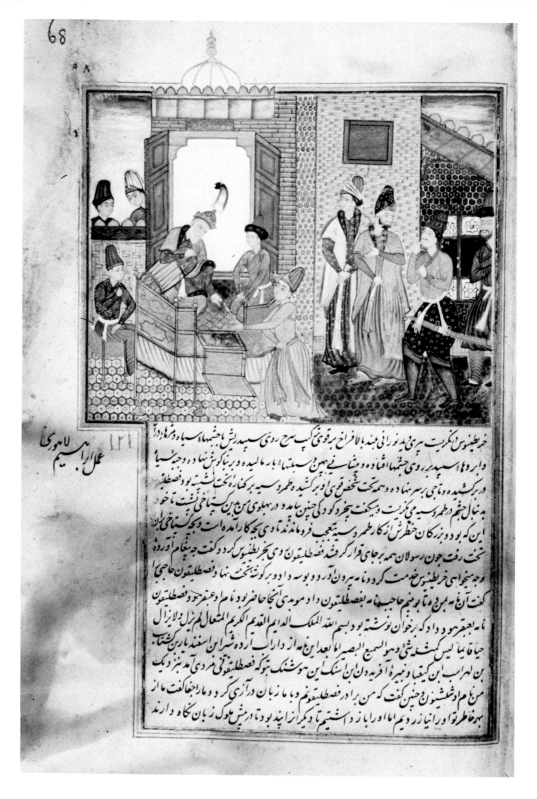

١٢١

عمل ابراهیم لاهوری

خربطینوس درنگریست پیری دید نورانی بلند بالا فراخ بر قوی تن کپ سرخ روی سپید ریش با چشمهای سیاه و مژهای دراز
وابروهای سپید بر روی چشمها افتاده و میشانی پهن و سبلتهای بار مالیده و بر بناگوش نهاده و جبه سیاه
در بر کشیده و تاجی بر سر نهاده و همه تحت تخص قوی او بر کشیده طمروسیه بر کناره تخت نشسته بود فصطلیقون
بدنبال چشم در طمروسیه می نگریست و میگفت بنگرد کودکی چنین بیامد و در پهلوی من بدین گستاخی نشست تا خود
این که بود و بزرگان خطرش از کار طمروسیه تعجب فرو ماندند تا وی بچه کار آمده است و بچه گستاخی بر آن
تخت رفت چون رسولان همه بر جای قرار گرفتند فصطلیقون روی بخربطینوس کرد و گفت چه پیغام آورده
وچه میخواهی خربطینوس خدمت کرد و نامه بیرون آورد و بوسه داد و بر گوشه تخت نهاد فصطلیقون حاجبی را
گفت آن نامه بمن ده تا بو بینم حاجب نامه بفصطلیقون داد موبدی آنجا حاضر بود نام او عبقر هود فصطلیقون
نامه بعبقر هود داد که بر خوان نوشته بود بسم الله الملک الدایم القدیم الکریم المتعال لم یزل ولا یزال
حیا قایما لیس کمثله شئ وهو السمیع البصیر اما بعد این نامه از داراب آرده شیر ابن اسفندیار بن گشتاسب
بن لهراسب ابن کیقباد و نبیره آفریدون ابن آتسنک این هوشنک بتو که فصطلیقونی مردی آمد بنزدیک
من نام او شلشیون چنین گفت که من برادر فصطلیقونم و با ما زبان درازی کرد و ما را جفا گفت ما از
بهر خاطر تو او را نیازردیم اما او را باز داشتیم تا دیگران را پند بود تا در پیش ملوک زبان نگاه دارند

سخن بغایت پسندیده نمود که بچگان خرس را از زنجیر رها کردند ایشان صورت زرگر را
صورت چوب دانستند فی الحال دردویدند و در دامن و آستین او خزیدن گرفتند زرگر
سر خند که ایشان را از خود می راندند ایشان لابه کنان می آویختند و منعم خود تصور میکردند قطعه
نخشبی یاد کن ز منعم خویش
نوش ناحق شناس نیش بود
کوشش منعم علیه همه
جانب منعمان خویش بود
حاکم گفت ای زرگر محقق شد که این بچگان تو اند و ایشان را ببر و سر در گریبان
بیان وقت خود کن و از کرده و گفته خویش مستغفر شو باشد که ایشان بر صورت

Pl. 48

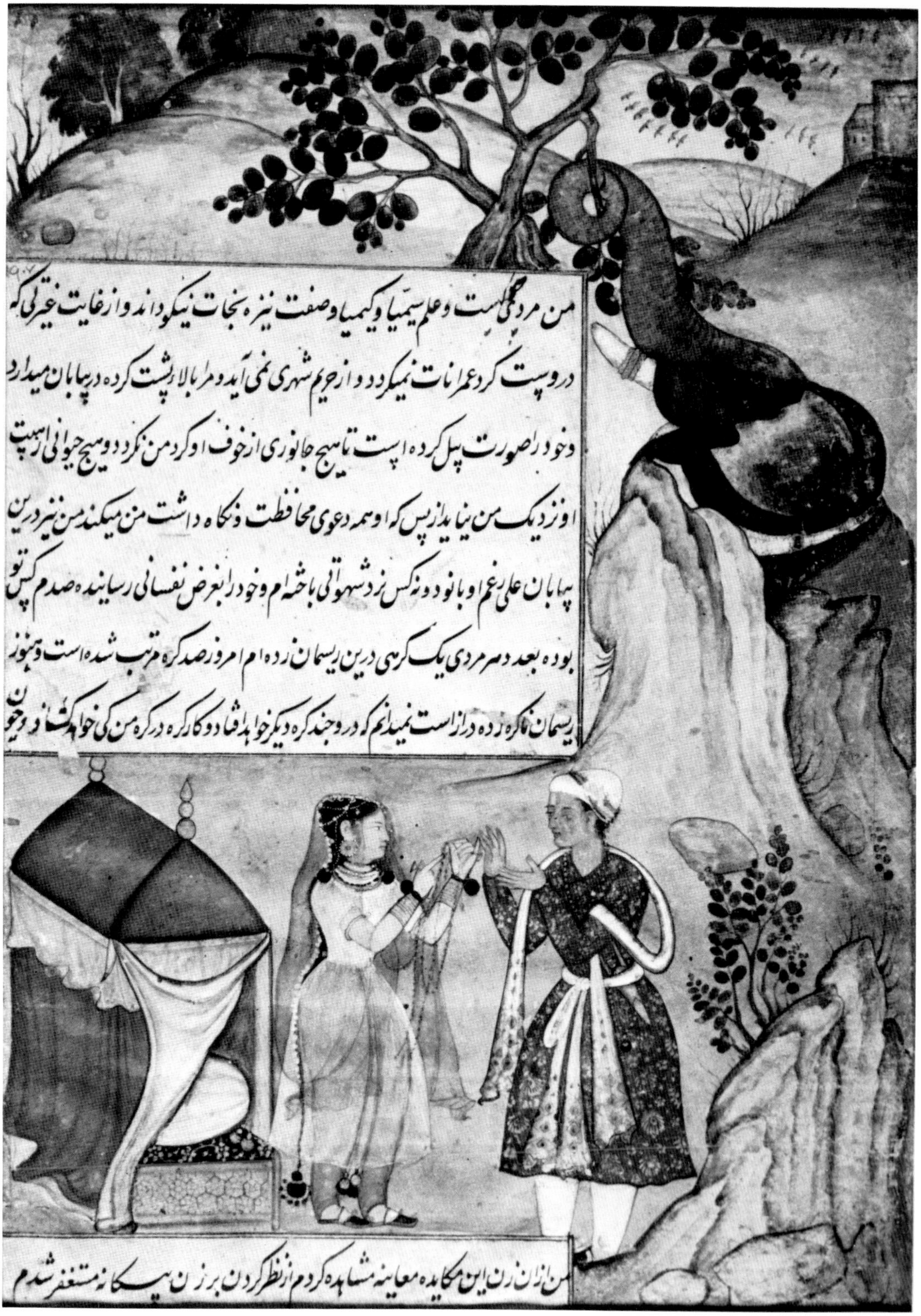

من مرد حکیمی است و علم سیمیا و کیمیا و صفت نیزه نجات نیکو داند و از غایت غیرتی که
در وی است گرد عمارات نمیگردد و از حریم شهری نمی آید و مرا بالا پشت کرده در بیابان میدارد
و خود را صورت پیل کرده است تا هیچ جانوری از خوف او گرد من نگردد و هیچ حیوانی از هیبت
او نزدیک من نیاید از پس که او همه دعوی محافظت و نگاه داشت من میکند من نیز درین
بیابان علی رغم او با نود و نه کس نرد شهوتی باخته ام و خود را بغرض نفسانی رسانیده صدم کس تو
بوده بعد هر مردی یک گرهی درین ریسمان زده ام امروز صد گره مرتب شده است و هنوز
ریسمان ناگره زده دراز است نمیدانم که در و چند گره دیگر خواهد افتاد و کار گره در گره من کی خواهد گشاد و چون
من از آن زن این مکاید معاینه مشاهده کردم از نظر کردن بر زن بیگانه مستغفر شدم

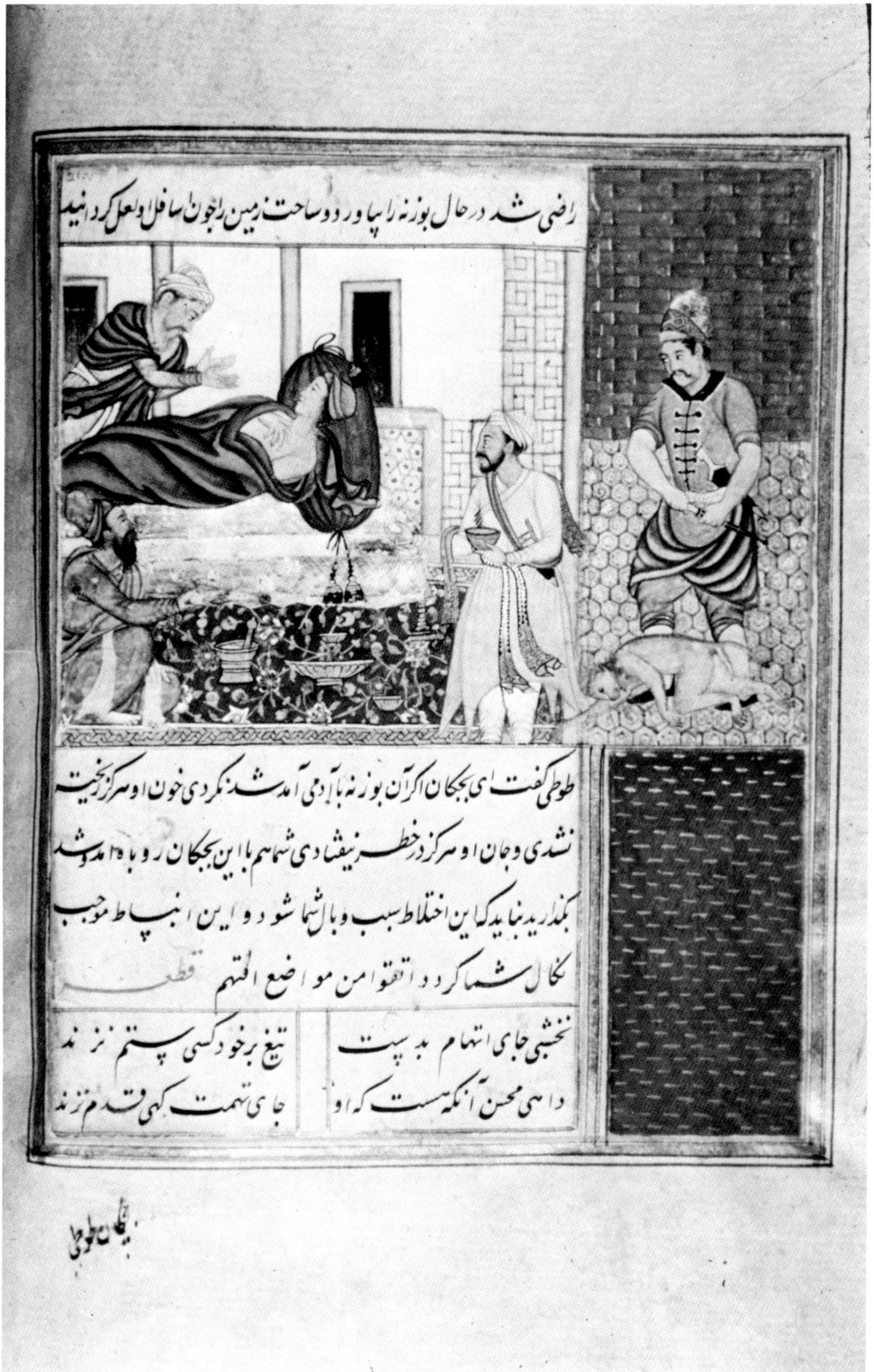
راضی شد در حال بوزنه را پاور دو ساحت زمین را چون سافل او لعل کردانید
طوطی گفت ای بچگان اگر آن بوزنه با آدمی آمد شد نکردی خون او هرگز ریخته
نشدی و جان او هرگز در خطر نیفتادی شما هم با این بچگان روباه آمد شد
بگذارید نباید که این اختلاط سبب وبال شما شود و این انبساط موجب
نکال شما گردد اتقوا من مواضع التهم
قطعه
نخشبی جای اتهام بد بپیست
داهی محسن آنکه هست که او
تیغ بر خود کسی چستم نزند
جای تهمت کهی قدم نزند
بچگان طوطی

هر هفت کس زیر آن درخت رفتند و آن عورت را نیز با خود ببردند و صورت
حال باز نمودند و حکم التماس کردند تنهٔ درخت بشکافت
و آن عروس پس درون رفت و در زمان مدد شد
و از هر برگ آن درخت آواز برمی برآمد کل شیٔ یرجع الی اصله
و آن هفت عاشق ازو خایب و خاسر و خاین شدند و با دلی در التهاب و دیده
در انسکاب بازگشتند و باقی عمر در حسرت و پشیمانی بگذرانیدند طوطی چون
سخن اینجا رسانید با خجسته آغاز کرد ای کدبانو بیم من آنست که ناگاه شوی تو

ن دران مجلس برفت شوی خود را دید صاحب مجلس ازان جا که حاضر جوانی او بود چادر
بسر بینداخت و دست در سر و ریش شوی بزد و فریاد برآورد
که ای مسلمانان مرا از دست این بی انصاف فریاد رسید مدت شش ماه است
که این جائی روان شده است و مرا دو دیده در راه مانده روزهاست

کودکی آن جوان ازان بست پیلی بساخت و درسید و بنهاد زن بعد از
لذت جسمانی سید بست بر سر کرده بر شوی خود رفت جون شوی پیل بست
تنغیر شد و کفت این جه استهزا رست که تو بر من میکنی و این چه کار
آنست که بدان مشغول می باشی زن جون این حال بدید بر فور آغاز

Pl. 54

آورد پادشاه خواست تا آنرا بخورد باز اندیشه کرد که هرچه
بخورده اند توان خوردم بر مزاج این میوه معلوم نیست و خاصیت
این ثمره مفهوم نه اول در حق خود چه امتحان کنم بعده پیر را بداد بمجرد
آنکه پیر آن بخورد جان از او وداع کرد پادشاه متغیر شد و متحیر گشت و گفت
عین مصلحت بود که من این میوه زهر نخوردم اگر من بگفت طوطی مغرور
شدمی و در خوردن آن میوه اقدام نمودمی همان معاینه کردمی که آن

باشي چون ميلي چند برفت آبي صعب پيش آمد جوان گفت من مردي ام
سياح در آشناگري دستي تمام دارم اگر بگويي اول نقدي كه هست آنرا گذار
برم بعده ترا گذارا كنم گفت نيكو باشد چون جامه و پيرايه او برگرفت و گذارا
شد در خاطر گذرانيده من مردي درويش و او دختر راي گذارا با پادشاه

ای خجسته عفت و صلاح همه وقت مطلوب است و عصمت و فلاح همه که
مرغوب اما هر چیزی را ایامی است و هر کاری را هنگامی ازبو عفت و صلاح
درین وقت همچنان هیچ نماید که ازان درازگوش از سرود گفتن هیچ نموده
بود خجسته پرسید آن چگونه بود طوطی گفت چنین گویند وقتی درازگوشی
بود و او با گوزنی محبت داشت و در مراتع یکجا بودندی و در مرابع یکجا غنودندی
شبی در اوان ربیع و هنگام بهار در باغی یکجا می چریدند ناگاه تلخه درازگوش بجنبید

Pl. 58

پر درون انداخت و از خضریات باغ دور شد بمجرد آنک خر در بانگ شد باغبان در
رسید و او را بگرفت و با درختی محکم بست و بدولت و شدت بکشید و دست
و پای او بشکست و گوشت و پوست او را خریطه کرد قطعه
تا رخ لعل تو نگردد زرد
هر که او گفت ناصحان نشنود
لت محکم زمانه نیکو خورد

مرا بشکست و صرصر تند خیز شوق بیخ هوش و عقل مرا برکند امشب مرا اجازت
ده تا سوی بیت الوصال دوست شوم و دیدهٔ تاریک را بنور حضور محبوب
منور گردانم طوطی گفت ای عذرا وقت و ای زلیخا زمانه اگر درین
وقت عذرا بودی از شرم خلق تو نام وامق نبردی و اگر زلیخا بودی از
خجالت قلقله تو قصهٔ یوسف نخواندی از جانب من ترا رخصت است برخیز
و جانب بیت الوصال محبوب شو و چون آنجا رسی شرایط خدمت بوفا رسان

کرد اند من از غایت غصه تیغ جانب وانداختم اما افسوس که دم او بریده شد و او زنده از پیش
من رفت زن رای را این سخن درو کرد و بخشم از پیش او برخاست مار چون این
سخن بشنید گفت لعنت بر آن ماده باد خود چه کرده است و با من چه گفت اکنون
نیکو شد که من این رای را خسته نکردم و زهر آن ماده برو نریختم مار از سبد بیرون
شد و رای را خدمت کرد رای گفت ای مار برای خوردن من آمده بودی این خدمت
چیست و اگر از برای خدمت آمده بودی خود بی اذن درون آمدن چه مار گفت
من جفت آن ماده ام که تو امروز دم او بریده و با من سبب بریدن دم نوع دیگر گفته بود

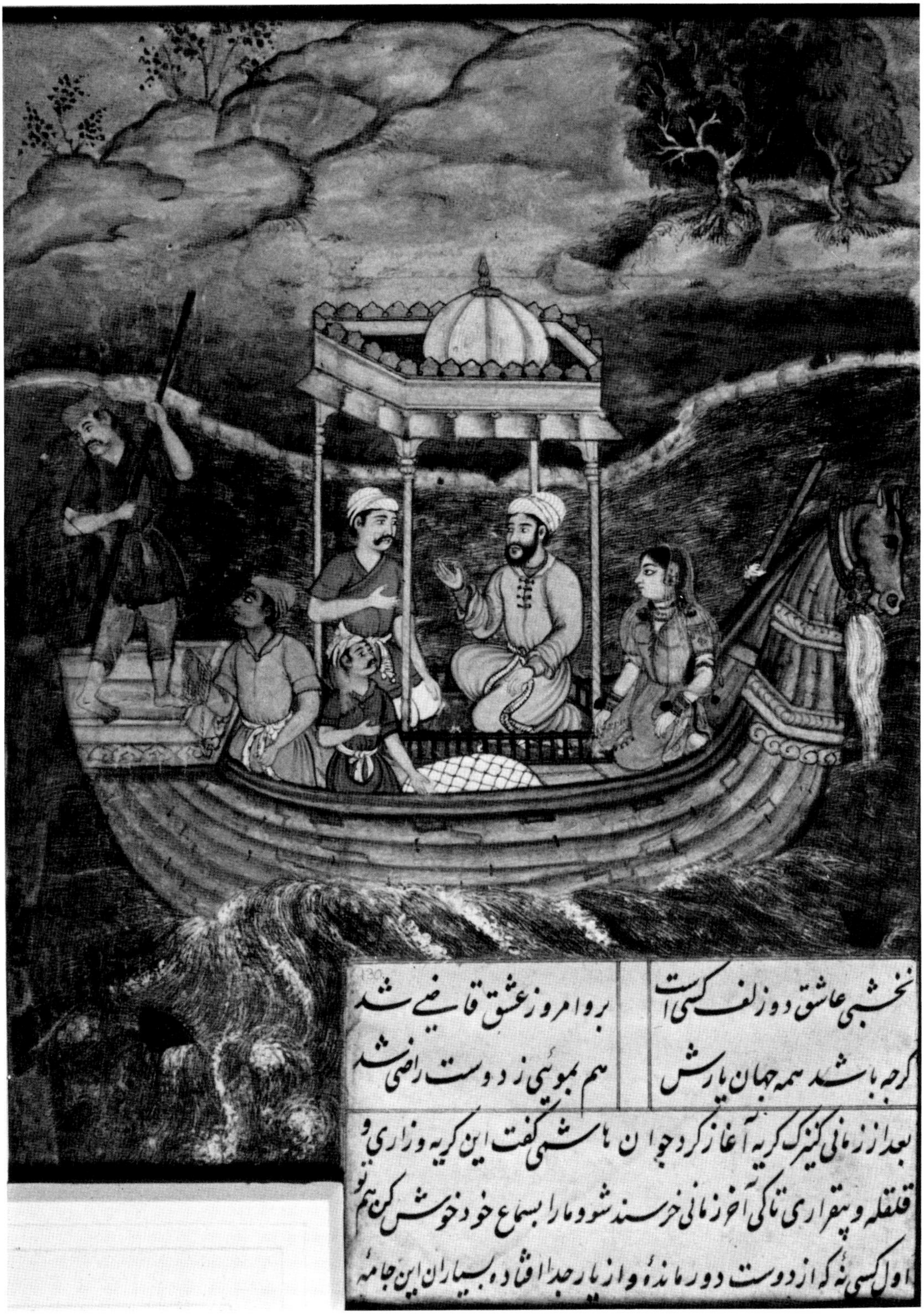
130
نخشبی عاشق دو زلف کسی است
برو امروز عشق قاضی شد
گرچه باشد همه جهان یارش
هم بموئی ز دوست راضی شد
بعد از زمانی کنیزک گریه آغاز کرد جوان هاشمی گفت این گریه و زاری و
قلقله و بیقراری تا کی آخر زمانی خرسند شو و ما را بسماع خود خوش کن هم تو
اول کسی نه که از دوست دور ماندهٔ و از یار جدا افتادهٔ بسیاران این جامه

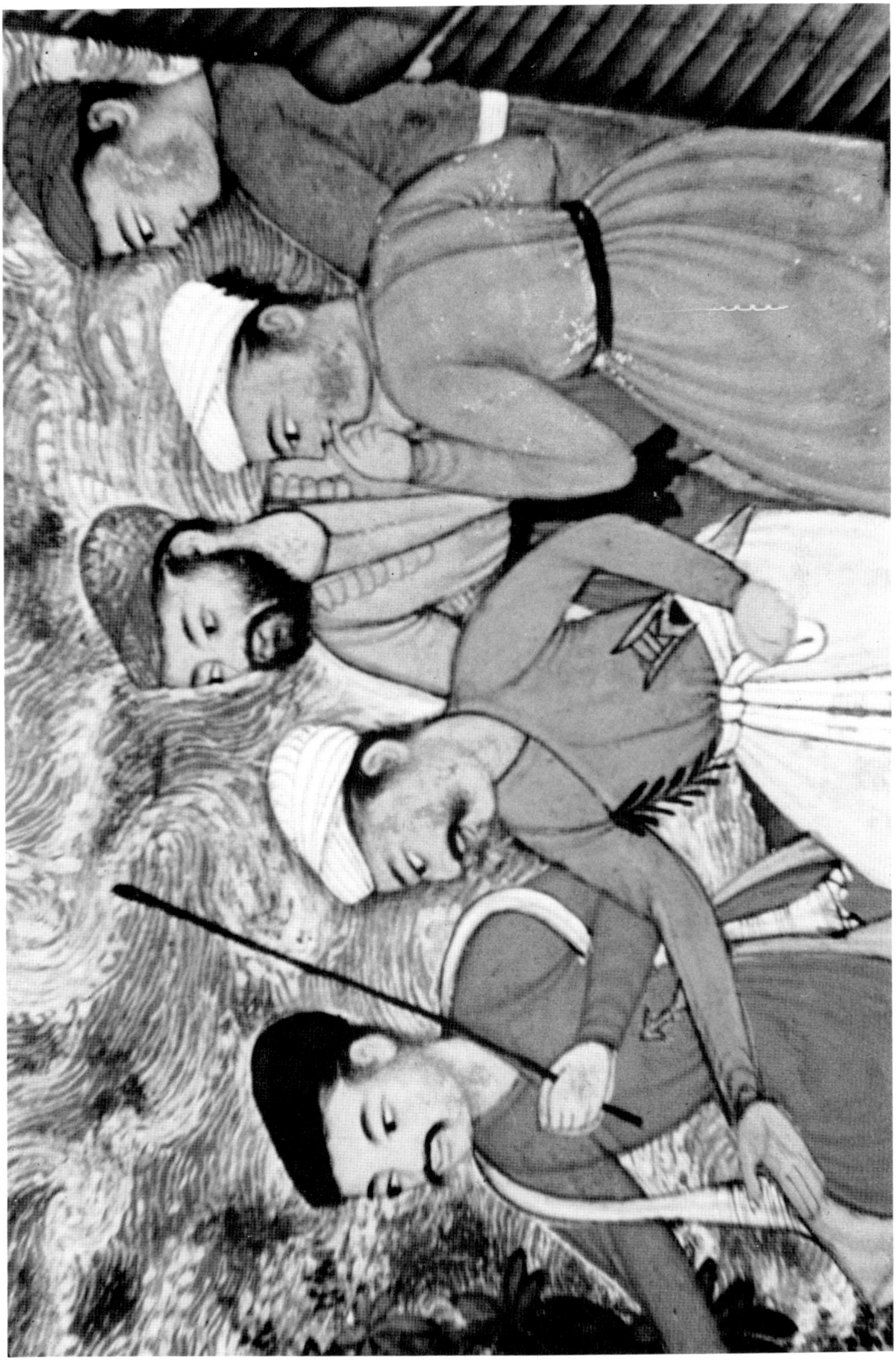

Pl. 66

Pl. 68

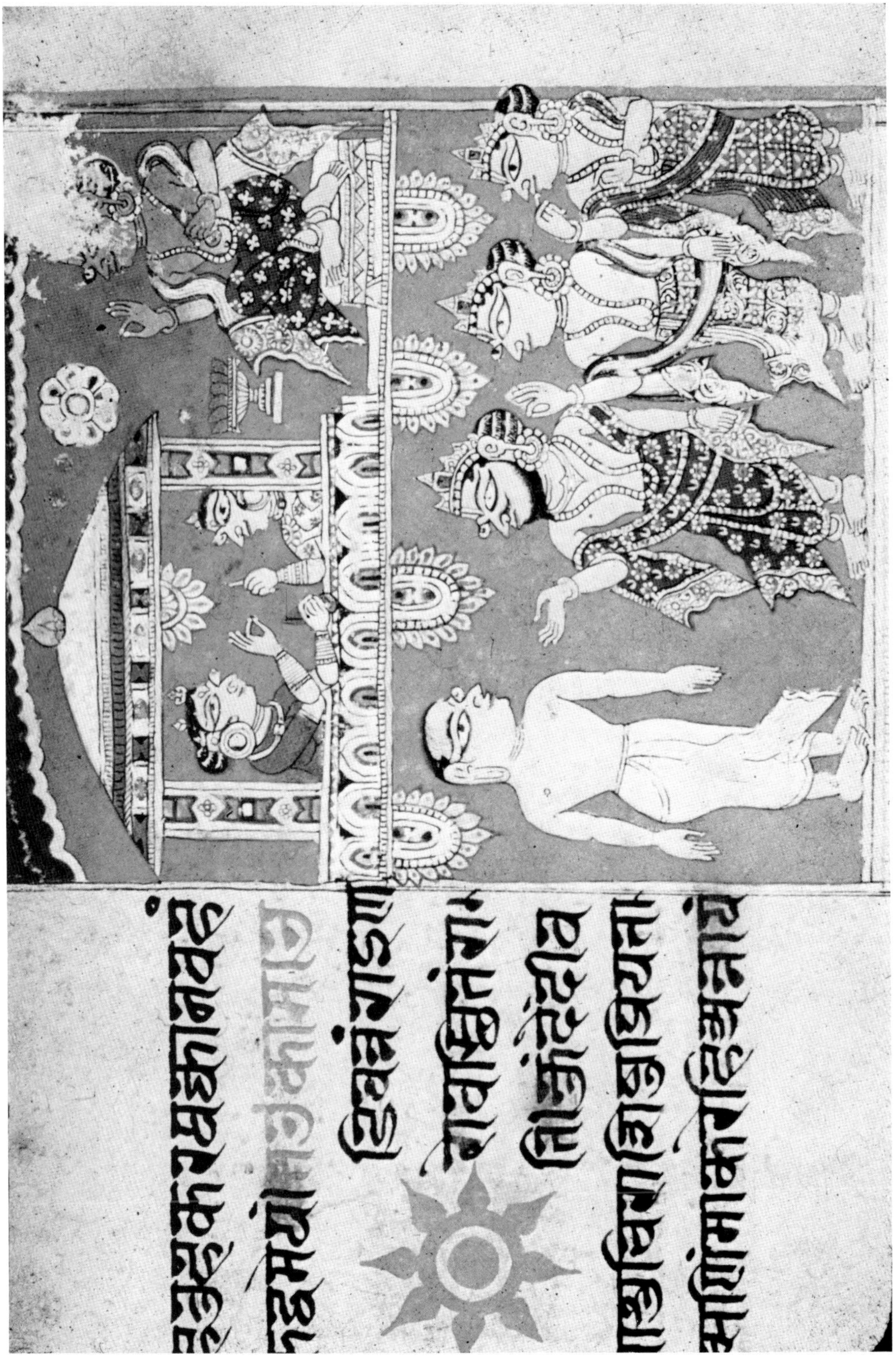

Pl. 74

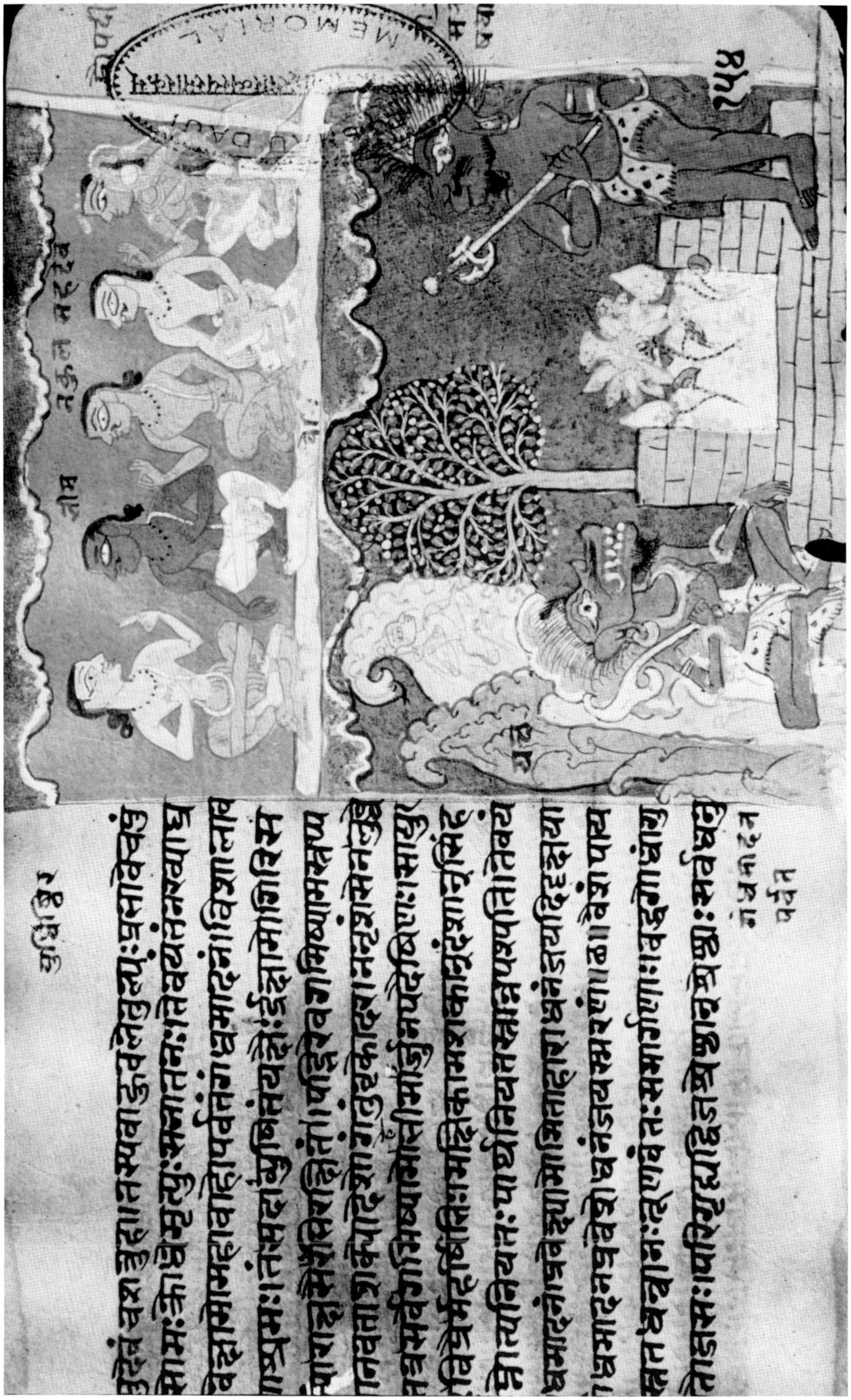

Pl. 76

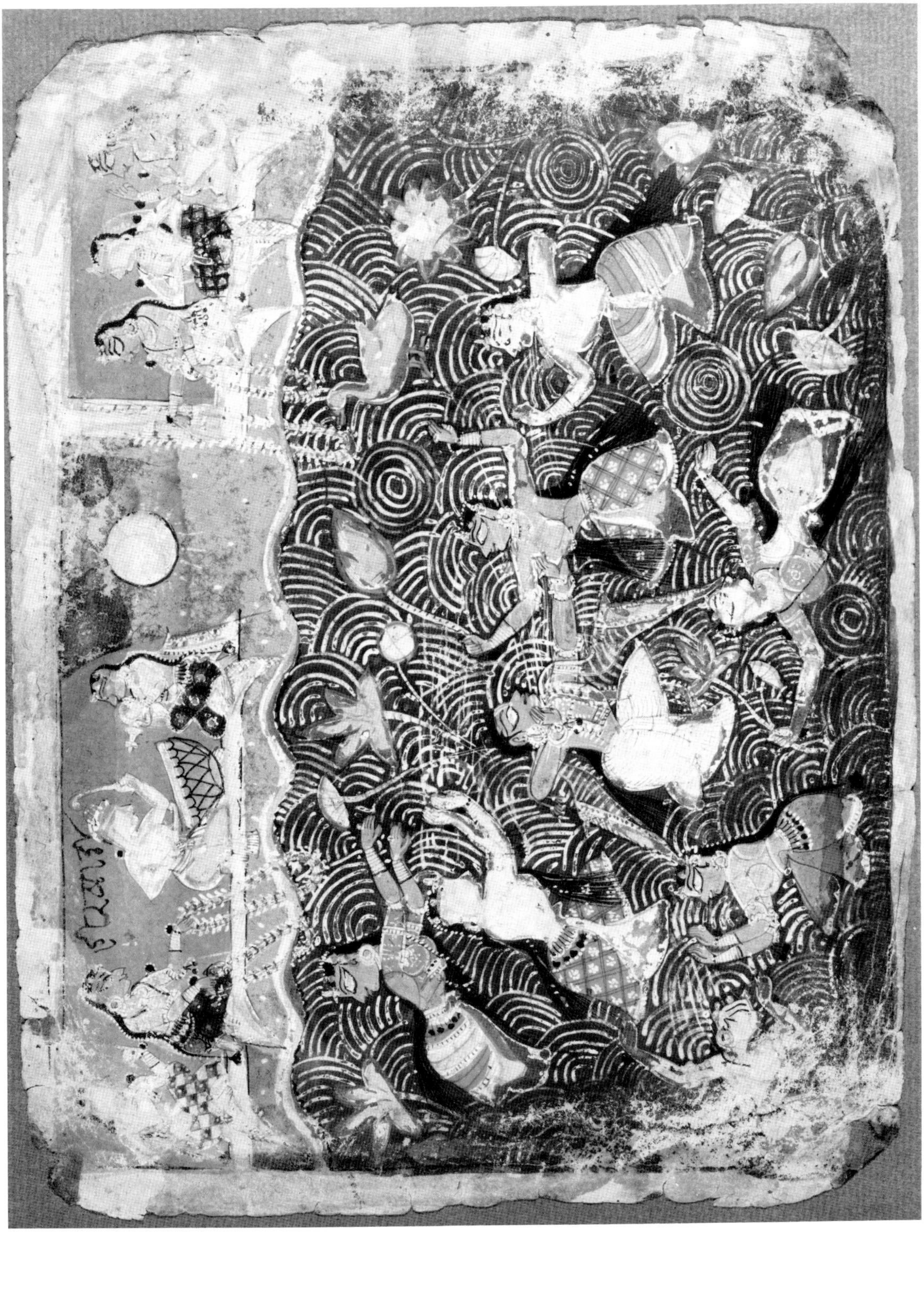

Pl. 80

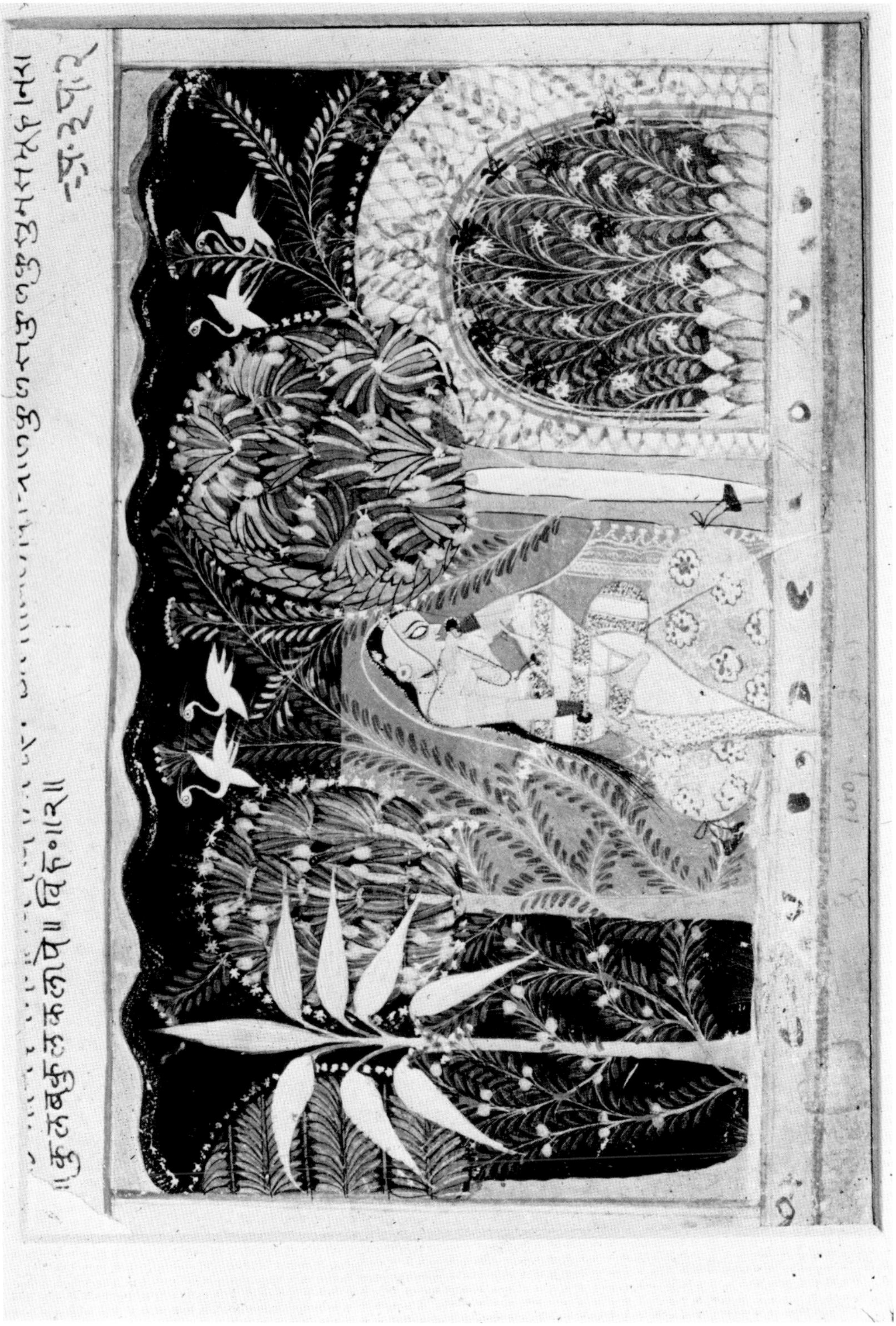

Pl. 81

Pl. 82

20

Pl. 84

Pl. 86

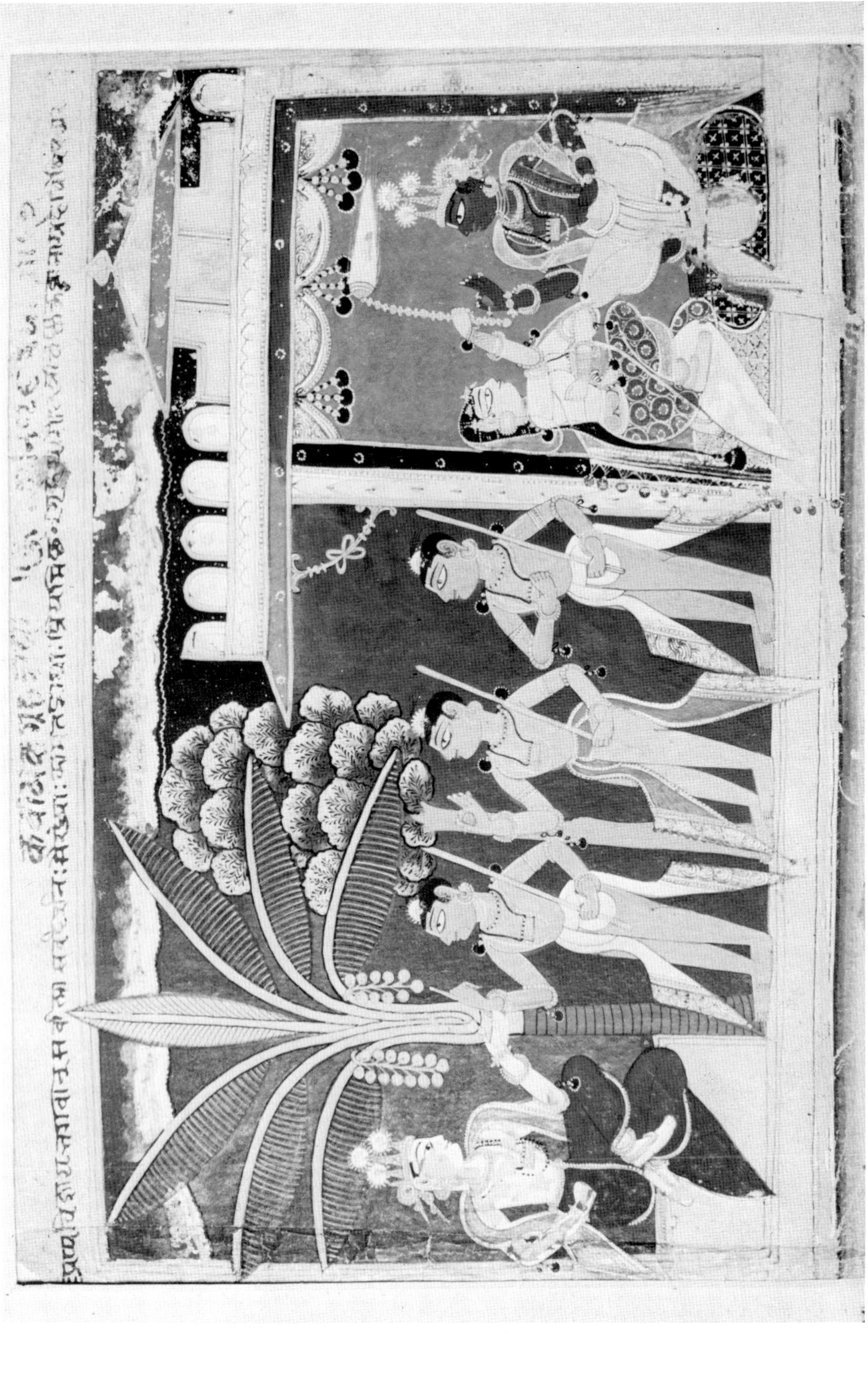

Pl. 90

मध्ये मतंगस्तु वसंतरागः॥ ॥२१॥ वसंतरागिणी॥ श्रीराग॥

Pl. 96

Pl. 98

Pl. 100

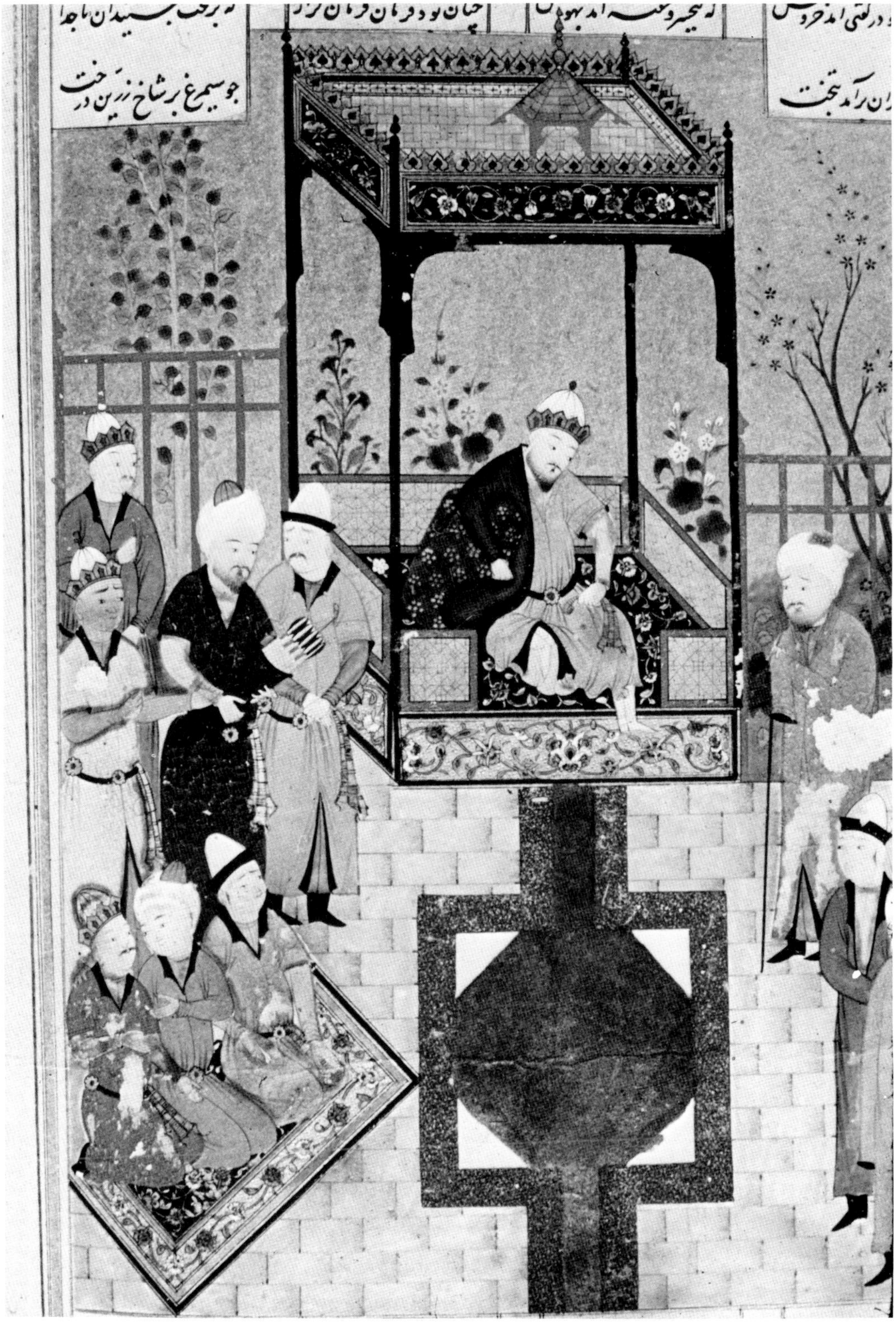

سرزک با سرهزان ودیا | بوا سدری سد بر سرهزر | نایش کنان کنت اکرتخت شا | کند بر سرتخت این بند راه | رود داور زمین یوس و

پذیرفت شه خواهش کرم | برفتن گکه داشت آزرم او | سرش با فسر کرامی کند | بدین سرنزرکیش نا

زمین ازسرکنج یکشا دبند | روار و بر آمد بجرنخ بلند | شه دنشکرشه پیکبار کی | بران خوان شده نداز ر

یکی تخت زد دید چون آفتاب | دروجشه درجود دریای آب | سکند رجو برخوان خاقان رسد | پی خضربرآب حیوان ر

شادی بران تخت زرین نشت | زکا فورو عنبر نزنجی بست

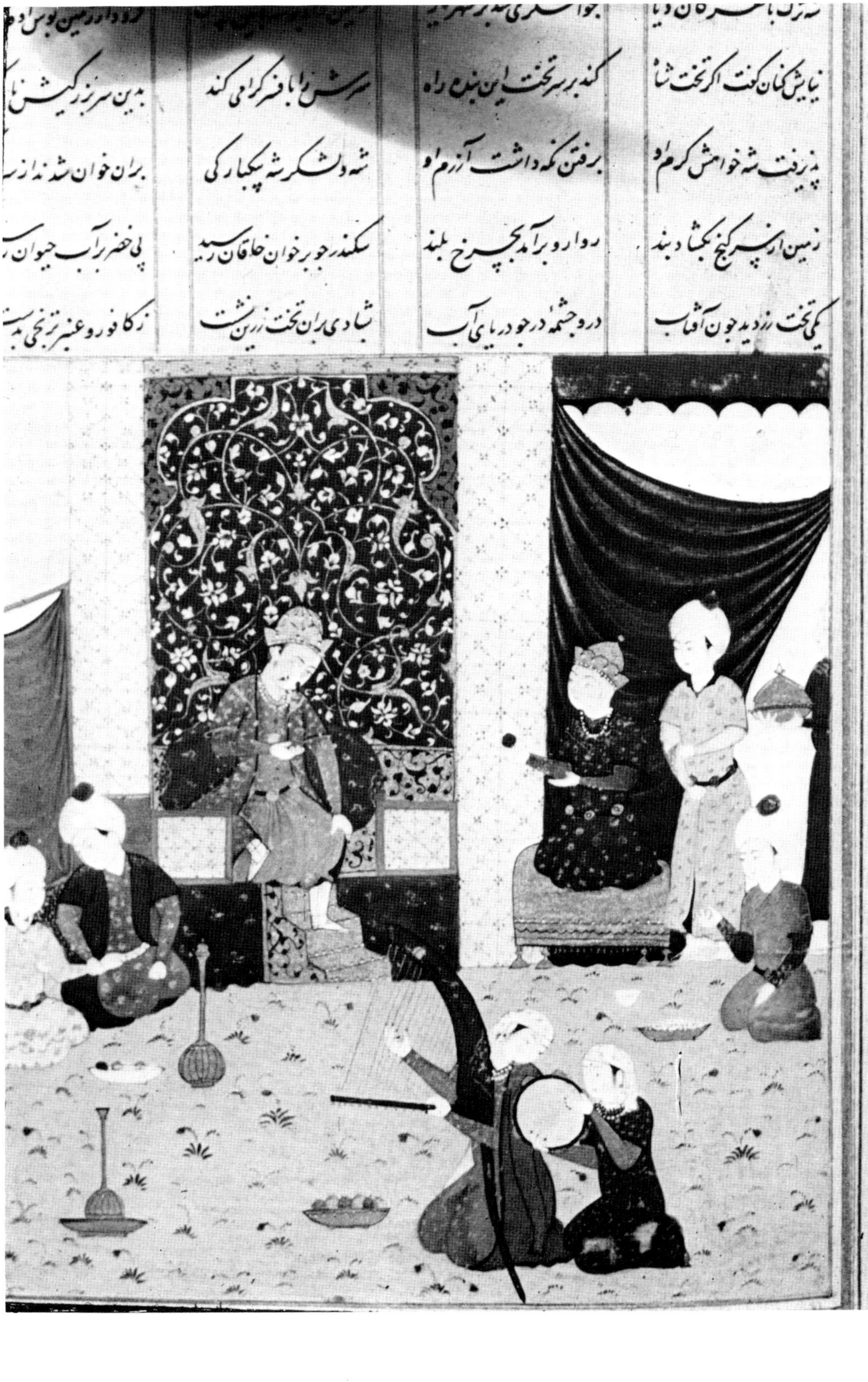

آما جکهی دوان از دور
زیر نخل منظور

جون تربت دوست در بر آورد
ای دوست بگفت وجان بر آورد

Pl. 106

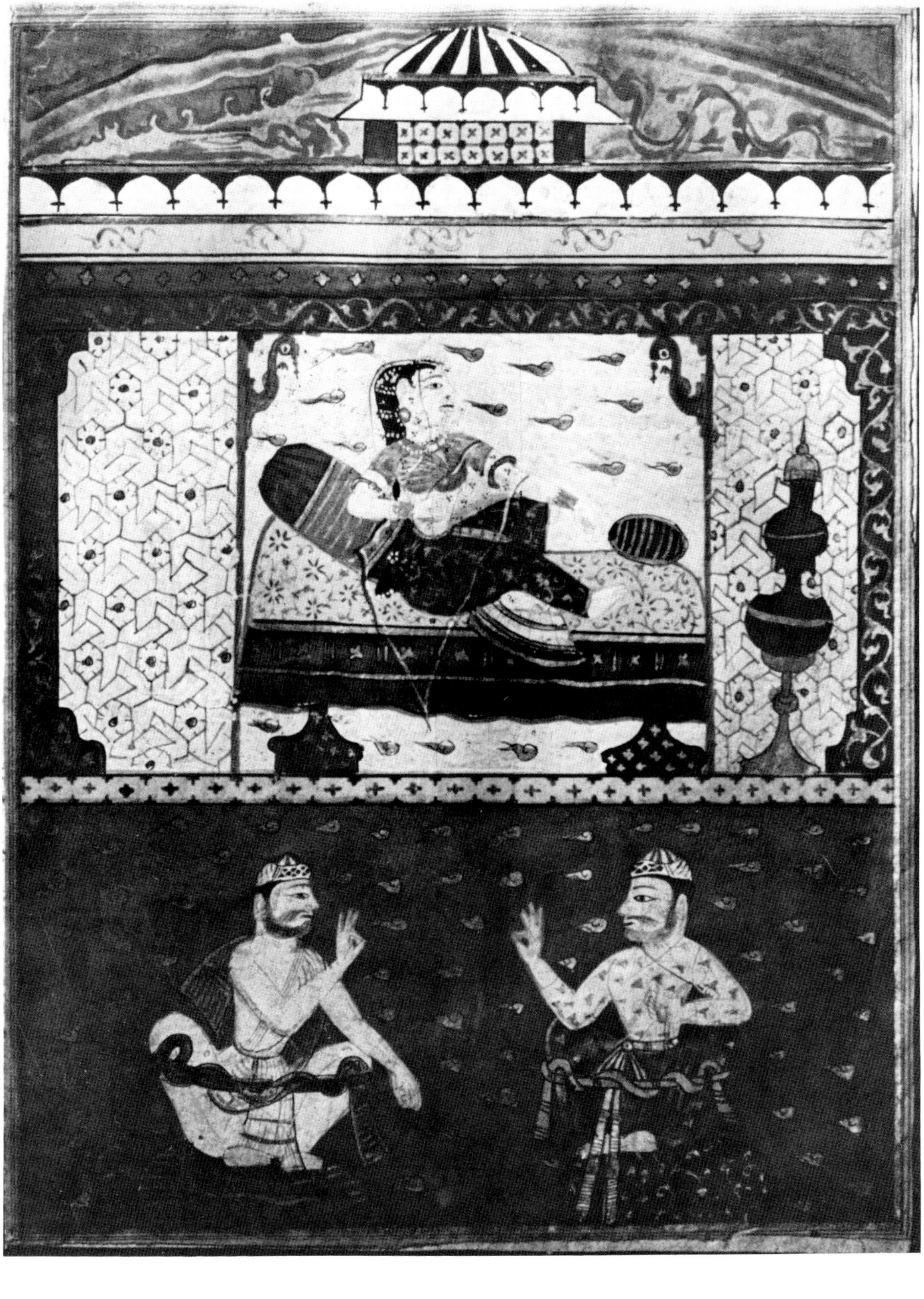

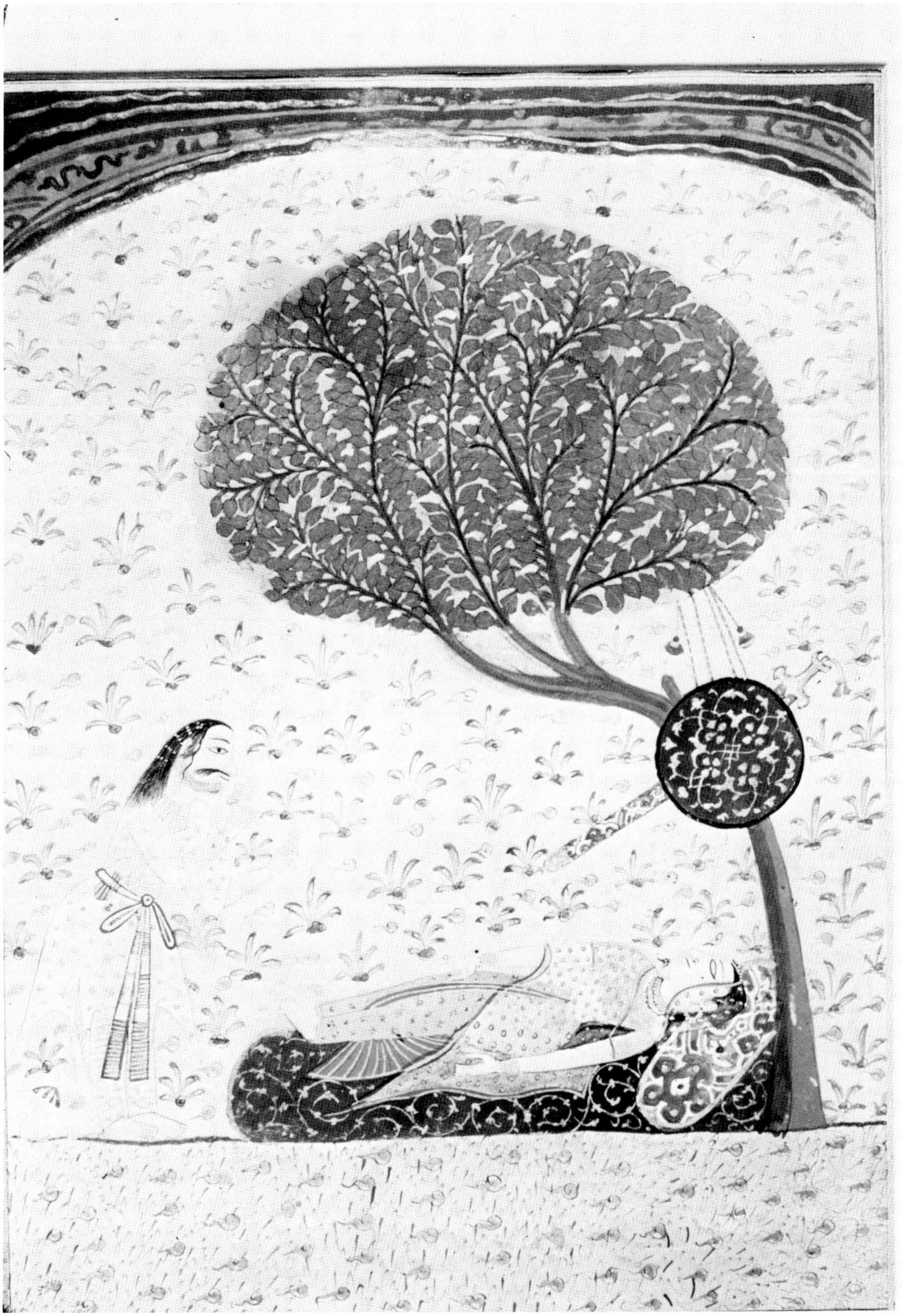

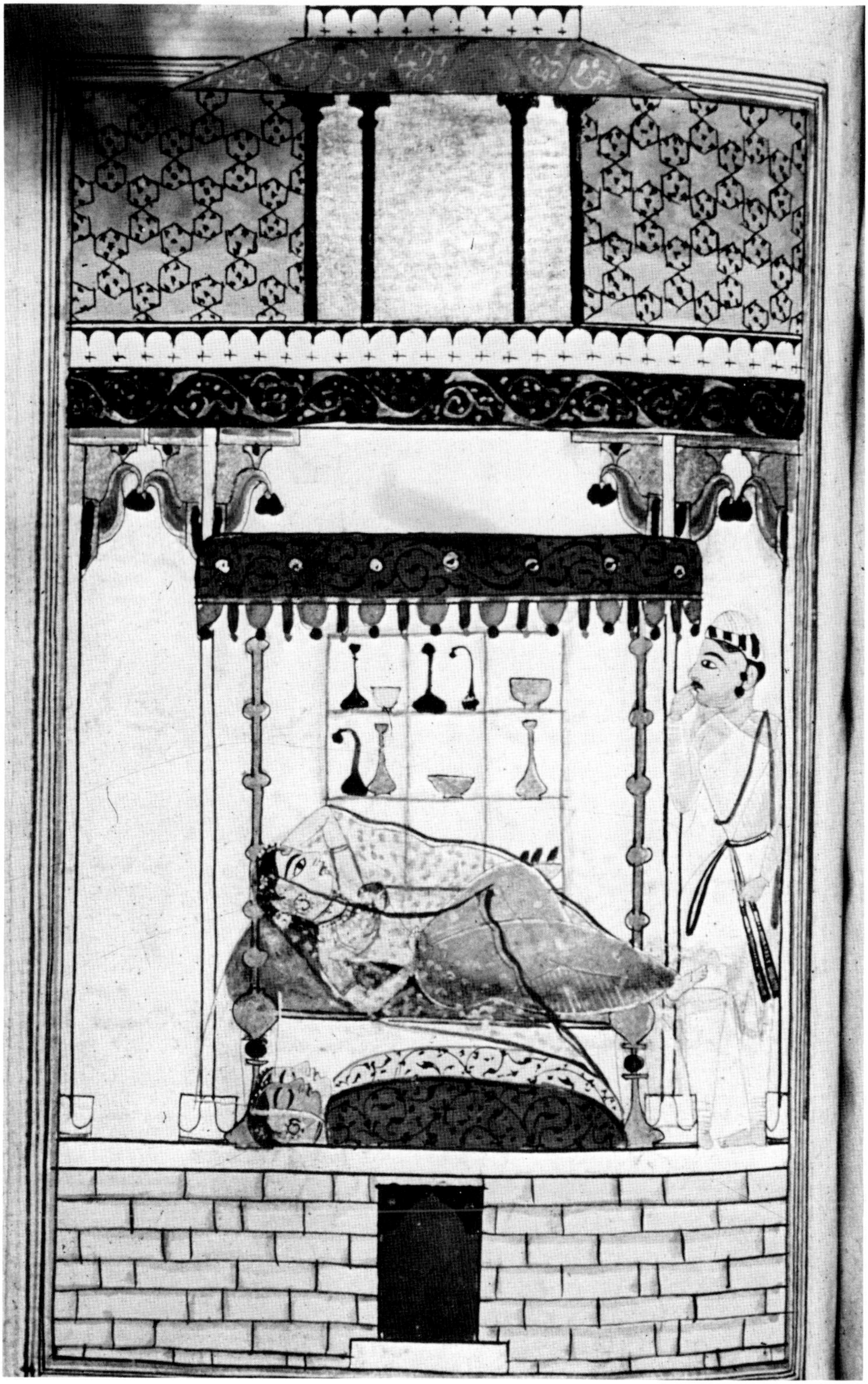

دریا بر صورت آدمی که آب لطافت از روی او میچکید و دُر و جواهر از دهن
آدمی بارید پیش رای آمد و از حال ورود و وصول او استفسار
کرد و گفت کجا رسیده و بچه مصلحت رنجه شده اگر حاجتی است بگو تا آن
مقتضی کنم و اگر مصلحتی هست اشارت کن تا گرد آن مصلحت برآیم رای
گفت مرا کاری سخت پیش آمده است و مهمی صعب معترض بحکم حیا و زملکا

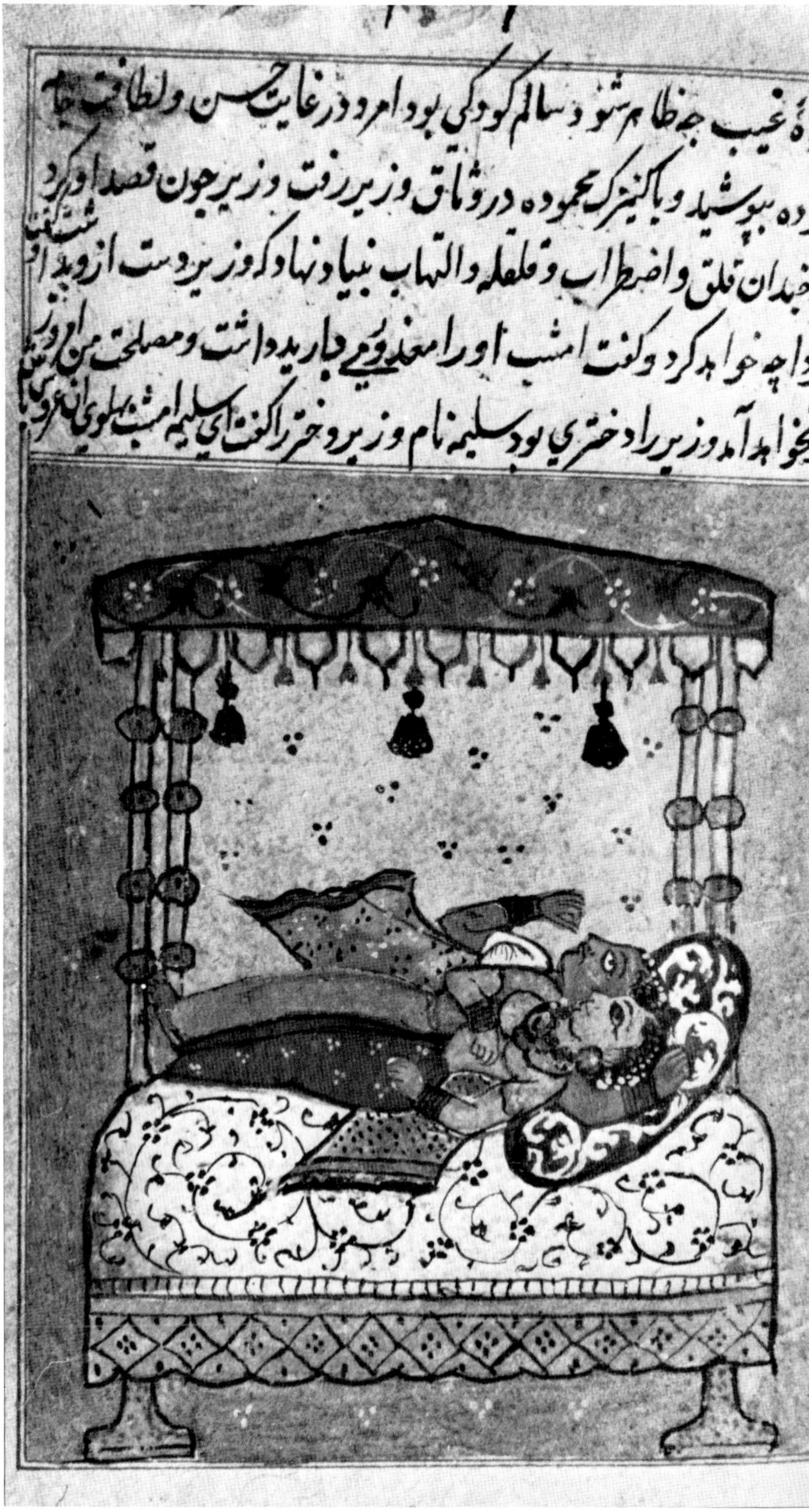
پرده غیب جه ظاهر شود سالم کودکی بود امرد در غایت حسن و لطافت
محموده بپوشید و باکنیزک محموده در وثاق وزیر رفت وزیر چون قصد او کرد
او خیدان قلق و اضطراب و قلقله و التهاب بنیاد نهاد که وزیر دست از وی بدا
فردا چه خواهد کرد و گفت امشب او را معذور می باید داشت و مصلحت من
بر نخواهد آمد وزیر را دختری بود سلیمه نام وزیر دختر را گفت ای سلیمه امشب پهلوی

مي بايد بود اکنون برخيز و جانب وثاق دوست شو خجسته خواست تا

همچنان کند روز از سياه کوشي با حزم تر بود در حال صبح چهره لمعاني

بکشاد و رفتن او در توقف افتاد

نخشبي خواست تا رود امشب — سوي خوبي که زد ز خوبي کوس

صبح از رفتنش بشد مانع — دشمن عاشقانست صبح خروس

داستان زنبه با اطفال او و حيله او در دادن پلنک را تخويف

پلنک و خلاص دادن زنبه خود را از پلنک شب سي ام

چون پلنک کرم مزاج افتاب از کوه سپهر در غار مغرب رفت

و ماه معتبد باطفال کثير از قماط مشرق برامد خجسته با باطني جوشان

و ظاهري خروشان بطلب رخصت بر طوطي رفت و کفت اي اشرف الکفاه

و اي اکمل الدهات تر اينکو محقق است که چهار چيز قابليت عود

ندارد قضا رفته و سخن کفته و تير انداخته و عمر کذشته عمري بدين

لطيفه لطيفي مراد در عزم ميکذرد و حياتي بدين نظيفي مراد راند وه

مرو

Pl. 116

جون بد ببيني نكو تو حيله كن اكر رود وقتي بد هم بيست

146

طوطي جون سخن اينجا رسانيد با خجسته اغاز كرد اي كدبانو ترا در خانه

از سياه كوش كه دوست نام كرده اند همبرين حزم و همبرين هوشياري

شهر ساخت و امر فرمود تا دهلی کهنه را خراب ساختند و مردم آنجا را باینجا آوردند و جلا

و جاها بنهای بلطف جمع

شهر همین شده و در اطراف آن باغات بسیار و بازارهای خوب پیدا شده و سلطان

سکندر پسر سلطان بهلول چون آگره را دار السلطنه ساخت دهلی بآن جمعیت نماند

و آگره شهری بزرگ شد و حضرت اعلی در آنجا قلعه ساخته‌اند که مثل آن هیچ دیده ور

ندیده و مانند آن در ایران و توران نیست و دهلی از اقلیم دوم است و مشایخ و علما

مقبره بسیاری از اکابر ضمه بوده اند

و اولیا در آنجا بسیاری از آنجاست و آن شهر معظم بغایت خرم و نزه است

و سلاطین هند در آنجا عمارت عالیه بسیار ساخته اند و زمین دهلی از سایر

بلاد هند بلطافت آب و هوا و فیض روحانی امتیاز تمام دارد امیر

خسرو دهلوی از آنجاست ... اصفهانی در ملازمت میرزا بدیع

الزمان حاکم سیستان می باشد سالک عشقی سیستانی است و خالی از کمالی نیست از اشعار

اوست   وه که گردون نمیکرد اجل شام فراق   او هم از من عار دارد زندگانی را ببین

با رقیبان سخنت در پس دیوار چه بود   با من دلشده این همه آزار چه بود

دی چو گفتی که بشمشیر جفایت بکشم   بر سر وعده رسیدی دگر انکار چه بود

جعفری را اگر از ناز بخواهی کشتن   این همه چشم زدن جانب اغیار چه بود

اسمش میر سید علی است و خلف صدق میر منصور مشهور است اصل ایشان

از ترمد است بعضی اوقات اجداد وی در بدخشان می بوده اند میر حیثیات بسیار دارد

و در وادی تصویر که امر موروثی اوست استاد بی نظیر است و در فن شعر و دریافت

آن بغایت بهره مند و خبیر میکند بیدار نوک خامه مشکین رقم

سر بسر صورت که آن مرقوم لوح فطرت است در بدو حال و ایام شباب در عراق

نشو و نما یافته در شهور سنه ست و خمسین و تسعمایه بکابل آمده بشرف ملازمت

حضرت جنت آشیانی سرافراز گشته و حضرت ایشان را بنا بر میل طبیعی بحیثیات

خصوصاً بتصویر التفات و توجه بسیار بجانب میر بوده تصویرات او را التفات

او را التفات میفرموده اند و حالا منظور انظار کیمیا آثار حضرت اعلی است [7]

2 صبحدم خار دم از همدمی گل میزد ناخنی در دل صد پاره بلبل میزد

این ابیات از سانح افکار بلاغت آثارش مرقوم افتاد

خواستم گویم از احوال خود آن بدخو را همدم همدم غیرست چه گویم او را

2 نیم بسمل صیدم و افتاده دور از کوی دوست میروم افتان و خیزان تا ببینم روی دوست

پیدرم از داغ سودای تو سر تا پای ماست تاجر عشقیم و اینها مایه سودای ماست

4 چون بنا کعبه ایست عشق بیابان او سبز نشانها کسان خار مغیلان او

قصاید و قطعات و رباعیات بسیار دارد همه مصنوع و مقبول مدت هفت

سال است که میر مذکور حسب الحکم حضرت اعلی در کتابخانه عالی بتحریر

و تصویر مجالس قصه امیر حمزه مشغول است و در اتمام آن کتاب بدیع انساب

که از مخترعات خاطر نقاد حضرت اعلی است اهتمام مینماید والحق آن کتابیست

که تا اوراق سپهر میناگون از تصویر کواکب ثواقب زیب و آرایش یافته نظیران

هیچ دیده ورندیده وتا اطلس سفینه گردون از حجره کشا هی ماه خورشید زینت ونمایش

گرفته دست تقدیر همچنان نسخه بر لوح خیال نکشیده و اختراع آن کتاب عجیب الابداع

برین وجه خیال فرموده اند که عجایب حالات و غرایب واقعات ان قصه را

از مبدا تا مال موبمو بر صحایف تصویر نگارند و از دقایق صورتگری دقیقه نامرعی

نگذارند وان حکایت در دوازده مجلد باتمام خواهد رسید هر جلدی مشتمل بر یکصد

ورق و هر ورقی محتوی بر دو مجلس تصویر بر صدر هر مجلس حالات و واقعاتی که یکدرع دریکدرع

بان صفحه متعلق است بزبان وقت املا کرده اند و بعبارت مرغوبی در سلک تحریر

درآورده وانشا و ابداع ان حکایات شوق انگیز و روایات طرب آمیز آن نسخه

اهتمام وسایح اقلام سحر انجام فصاحت شعار بلاغت دثار کمالات اثار خواجه عطاالله

منشی قزوینی که طبع وقادش مایده کلمات دلفریب است صورت انجام واتمام می یابد

و با آنکه در مدت مذکور سی نفر از مصوران نادره وصنعت مانی سیرت دران کتاب بر دوام

باهتمام کار میکنند زیاده از چهار جلد باتمام نرسیده کمال زینت ونهایت پرکاری

آن ازین معنی قیاس توان کرد وفقهم الله باتمامه فی ظل دولة العالیه وایالته جذبی

بادشاه قلی نام دارد جوانی جوان سپاهی زاده شجاع مستعد است بحسن خلق

و ملایمت طبع و ذکا متصف است خلف صدق شاه قلی خان نارنجی است

اصل ایشان از کردستان حوالی بغداد است پدرش نیز حسن طبع دارد مردم بادر ست

جذبی را طبیعت بشعر مناسب افتاده از اشعار اوست

در اوایل [illegible] مذکور [illegible]
باستاد [illegible] الملک خواجه عبدالصمد مصور شیرازی
[illegible] خواجه [illegible] آن [illegible]
اهتمام [illegible] آورده [illegible] آن کتاب [illegible]

در حجاب احتجاب نماند که در ترکیب احجام این عرایس و ترتیب اختراع این نفایس
که بهر ورقش ابواب دقایق و لطایف بر ارباب حقایق و معارف کشوده و بهر صفحهٔ
مطلعی چون مهر سپهر از عالم خیال طلوع نموده با وجود کثرت ترددات و اسفار و التزام
خدمات در لیل و نهار بتقصیر از خود راضی نکشته هر نکته که از شهسوار فکر در خاطر فاتر
کذشته بقلم حقایق رقم نوشته و در تتبع آثار و اخبار بدل مجهود و جهد موصول نموده
تا ابواب فتوحات معانی بمفاتیح توفیق ربانی کشوده تالیفی چنین حاوی غرایب حالات
و عجایب واقعات و جمع اشعار دلاویز و حکایات و روایات شوق انگیز و شرح
بلاد و کیفیت احوال سلاطین عالی نژاد و اقطاب و اوتاد بمستشهدات آیات
و امثال نموده ۞ دلم چه مایه جکر خورد تا بدانستم ۞ فلان چه کفت و چه سان بود و چون شد از دنیا
ابتدای جمع این کلدستهٔ معنبر که مشام ایام از شمیم نسیم آن معطر است در دیباچه بتوریت
اسم کتاب اشارتی بآن نموده و اختتام این مجموعهٔ محسنات جان پرور که ارباب ذوق را
چون جان شیرین درخور است در ماهی بود که تاریخ آن از این ابیات معلوم و کشوده میکردد

اللهم صل علی محمد نبی المرسلین

هذا الکتاب نتایج آثار ما سلف ۞ فانظر بجده نادر در من الترف
للمولف من بحر فکره ۞ بل اوجد المصنف من فکر بکره
ب العقل من لدیه ۞ قد قال بالبدیهة تمت علی یدیه
۹۷۹

تمت

کتب بعض اجزاء الکتاب هذا العبد الفقیر الحقیر نظام احمد دهلوی
فی تاریخ ثامن ربیع الثانی سنه ثمان و
تسعمایه بظاهر بلده اجمیر

INDEX

For brevity in the sub-entries, the following abbreviations have at times been used: ACB, A. Chester Beatty Library, Dublin; BKB, Bharat Kala Bhavan, Varanasi; BM, British Museum, London; CMA, Cleveland Museum of Art; Lalbhai, Kasturbhai Lalbhai collection, Ahmedabad; LDI, Lalbhai Dalpatbhai Institute, Ahmedabad; NMI, National Museum, New Delhi; POW, Prince of Wales Museum of Western India, Bombay; V and A, Victoria and Albert Museum, London.